IET COMPUTING SERIES 23

Data as Infrastructure for Smart Cities

IET Book Series on Big Data – Call for Authors

Editor-in-Chief: Professor Albert Y. Zomaya, University of Sydney, Australia

The topic of Big Data has emerged as a revolutionary theme that cuts across many technologies and application domains. This new Book Series brings together topics within the myriad research activities in many areas that analyze, compute, store, manage and transport massive amounts of data, such as algorithm design, data mining and search, processor architectures, databases, infrastructure development, service and data discovery, networking and mobile computing, cloud computing, high-performance computing, privacy and security, storage, and visualization.

Topics considered include (but not restricted to) IoT and Internet computing; Cloud Computing; Peer to Peer computing; Autonomic computing; Data centre computing; Multi-core and many core computing; Parallel, distributed and high-performance computing; Scalable databases; Mobile computing and sensor networking; Green computing; Service computing; Networking infrastructures; Cyberinfrastructures; e-Science; Smart Cities; Analytics and data mining; Big Data applications; and more.

Proposals for coherently integrated International co-edited or co-authored handbooks and research monographs will be considered for this Book Series. Each proposal will be reviewed by the Editor-in-chief and some board members, with additional external reviews from independent reviewers. Please email your book proposal for the IET Book Series on Big Data to: Professor Albert Y. Zomaya at albert.zomaya@sydney.edu.au or to the IET at author_support@theiet.org.

Data as Infrastructure for Smart Cities

Larissa Suzuki and Anthony Finkelstein

The Institution of Engineering and Technology

Published by The Institution of Engineering and Technology, London, United Kingdom

The Institution of Engineering and Technology is registered as a Charity in England & Wales (no. 211014) and Scotland (no. SC038698).

The Institution of Engineering and Technology
Michael Faraday House
Six Hills Way, Stevenage
Herts, SG1 2AY, United Kingdom

www.theiet.org

British Library Cataloguing in Publication Data
A catalogue record for this product is available from the British Library

ISBN 978-1-78561-599-3 (hardback)
ISBN 978-1-78561-600-6 (PDF)

Typeset in India by MPS Limited
Printed in the UK by CPI Group (UK) Ltd, Croydon

Though women have been pivotal in the creation of astonishing innovations and modern technologies, their story is not one that's often told, given credit to nor celebrated.

We dedicate this book to all the women in science, technology and engineering who have built, and continue to build, the foundation of modern programming and the Internet, unveiled the structure of the DNA and revolutionized the communication industry, broke codes to win wars and fought to remove the gender barriers to create the solutions that have changed the world.

Contents

About the authors

Dr Larissa Suzuki (PhD, FRSA, AFHEA, MIET) is an award-winning computer scientist, inventor and entrepreneur. Her career includes over 11 years advancing many fields of computer science and engineering, including smart cities, data infrastructures, emerging technology, AI, and computing applied to medicine and operations research. Her continuing academic work is based at UCL where she serves as an Honorary Research Associate in Computer Science. Her current industry appointment is at Oracle, where she is a Senior Product Manager for Automatic Machine Learning. She has published several research papers in leading academic journals, books and conferences; she is a frequent Keynote, conference, panel and tutorial speaker. She is the chair and founder of the Tech London Advocates Smart Cities, serves as a judge and reviewer of the ACM Research competition, and is a reviewer of scientific journals of the IEEE and Springer.

Prof Anthony Finkelstein holds a Chair in Software Systems Engineering at University College London, UK, and a senior appointment in HM Government. His continuing academic work is based at The Alan Turing Institute (ATI). He is on the Council of EPSRC and the Board of the ATI. He is a Fellow of the Royal Academy of Engineering and was awarded a CBE in the Queen's Birthday Honours list in 2016.

Chapter 1
Introduction

In the past four years, organizations, citizens and research institutes have become adept at holding governments to account for the environmental, social and economic consequences of population growth [1–3]. Since then, cities have been ranked on the basis of the "smartness" of their smart cities strategy, data and technological infrastructures (e.g. [4], Global Open Data Index[1]), and despite sometimes divergent and questionable methodologies, these rankings attract considerable publicity and marketing of technology products. As a result, smart cities have emerged as an inescapable priority for policymakers in every city in the world.

Every city competing in the smart cities arena has a digital strategy, which may have been developed through a planning process or it may have evolved through the ongoing public services offering. Each functional department and external organisations providing public services pursue approaches dictated by its professional orientation and the incentives of those leading the smart cities vision. Consequently, either a top-down or bottom-up approach is assumed to be the best approach to follow. However, the adoption of only one of these approaches alone rarely equals the best strategy. In many cases, such approaches have been susceptible to failure from inadequate stakeholder input, requirements neglecting, and information fragmentation and overload. They are also likely to be limited in terms of both scalability and future proofing against technological, commercial and legislative change.

The emphasis being placed on digital strategy planning in cities across the world reflects the proposition that there are significant benefits to gain through a methodical process of formulating strategy, to ensure that the activities and processes of cities and their partners are coordinated and directed at the same goals and shared key performance indicators. Increased attention to formal digital strategy planning has highlighted questions that have long been of concern to city leaders and policymakers: What are the important governmental, social and political factors presenting opportunities for new businesses and services? What public value my citizens are expecting from the city, and what is the best way to respond? How will my technology portfolio evolve and be sustained in the long run?

This book takes smart cities design from a broad vision to a city-wide collaborative and consistent configuration of activities. Its purpose is to improve the quality

---

[1] https://index.okfn.org/

of cross-domain city service management through improving the selection and definition of data infrastructures, speeding decision-making process, readying the city about exploiting data and digital services and informing implementation and data infrastructures and service operations.

This new form of design disaggregates a smart city into processes so that cities can use to think strategically about how systems, businesses and individuals can draw effectively on interoperable cross-domain city data and technologies. Data as infrastructure for smart cities provides for the first time the tools to strategically enable processes by which innovative use of technology and data coupled with governance strategies; a strong value network of partners can help deliver the various visions of data strategies for cities in more efficient, aligned and effective ways. Cities can take advantage of unprecedented insights into how the city and its infrastructure functions and be ready to overcome social challenges. The framework presented in this book has guided the design of several urban platforms in the European Union[2] and the design of the City Data Strategy of the Mayor of London.[3]

This book offers a rich framework for understanding the underlying inside out and outside in forces influencing the sustaining of data infrastructures and strategies. The framework reveals the important differences among stakeholders, how data infrastructure evolves and helps cities to find a unique position in realising their smart cities vision. It provides tools for capturing the richness and heterogeneity of data, services and technologies while providing a disciplined structure for examining them. This book brings structure to the concept of business strategies in smart cities by defining it in terms of urban capabilities and linking it directly to smart cities vision.

This book also signals a new direction and provides an incentive for data infrastructure design. The prevailing approaches to the design of smart cities were mainly driven by industry push, which were mainly concerned about technology deployments. Cities focused mainly on technologies, and technology developments in cities were presumed equal or differing primarily in size or in unexplained differences in strategies. The prevailing view of designing smart cities was a top-down approach in which cities accepted technology push without understanding its need and blueprint. Decision makers are often absent from a systematic view of how linking their smart cities vision, societal needs and technology developments. Moreover, they also lacked the tools to model and reason about data infrastructures design.

We present a comprehensive framework of design techniques to help decision makers in cities analyse its business strategies for strategies as a whole and design data infrastructures to support these activities, to understand its stakeholders and their expectations and to translate this analysis into a competitive strategy for data infrastructures. The content includes design templates, service selection methods, data profiling, gap analysis, key issues in data infrastructures design and means of resolution, examples and case studies.

---

[2] https://eu-smartcities.eu/sites/default/files/2017-09/EIP_RequirementsSpecificationGLA_%20V2-5.pdf
[3] https://files.datapress.com/london/dataset/data-for-london-a-city-data-strategy/2016-05-19T15:39:34/London%20City%20Data%20Strategy%20March%202016.pdf

Decision makers looking for concrete ways to tackle data infrastructure design will find answers to difficult questions in this book. To analyse and design a data infrastructure for the smart city, the reader can draw on the book in several ways.

Part I of this book lays the foundation for the development of data infrastructures and offers an introduction to the concept of smart cities, current challenges and urban capabilities. We discuss the emerging technologies and data requirements in smart cities, the transition from open data catalogues to data infrastructure maturity, both declining and new approaches to the management of city data and the key drivers and enablers for data infrastructures design and development.

Part II presents a general framework for analysing the structures of a data infrastructure and its business models. The underpinning of this framework is the analysis of the five main design components driving the dynamic business models of data infrastructures, namely, services, technology, organisation, value and governance. This framework is the starting point from which much of the subsequent discussion in the book begins.

This chapter builds on this dynamic framework to present techniques for the analysis of use cases, techniques for identifying stakeholders and their needs and expectations, city data from suppliers to consumers and techniques to model its logistical distribution, concepts for making and responding to demand for services, value design, and an approach to mapping strategic stakeholders' groups in a data, technology and digital services supply chain and explaining differences in roles and influence on the data infrastructure. The business models design serve as an input to a data infrastructure's value chain, which is designed and managed to explicitly consider the activities and processes that enable the stakeholders of smart cities to efficiently leverage their collective knowledge at R&D, Procurement, Roll-out and Market stages. We also discuss some of the major external forces affecting the development of data infrastructure throughout its life cycle.

Part III shows how the framework described in Chapter II can be used to develop data infrastructures and its business models, in particular, important types of city and government sector environments. These differing environments are crucial in determining the strategic context in which a data infrastructure is designed, the value network of partners providing an integrated approach to city data management, the strategic technology alternatives available and the common strategic errors in designing data infrastructures. This part also presents real-world examples of data infrastructure design and important types of strategic decisions that decision makers must make: vertical integration, major data services expansion and entry into the data infrastructure market. The analysis of each strategic decision draws on the application of the general frameworks of Parts II and III as well as on other theories and introduction on considerations in managing cross-domain data and digital services discussed in Part I.

Part IV of this book completes the business models' framework by systematically examining the viability of the designed data infrastructure model that can be assessed based on understanding the relationship between the data infrastructure's critical design issues and critical success factors. We demonstrate how to assess the viability of business models using the case studies presented in Part III. It concludes the discussion of the dynamic framework by examining ways of predicting the process of data infrastructure evolution and some of the implications of that evolution of a smart cities strategy. Part IV is designed not only to help a city make these key decisions but also to give it insight into how its value network, business models, stakeholders of services, data providers and suppliers, and technologies are joined together to deliver data infrastructures.

The parts of the book are meant to enrich and reinforce each other. Sections seemingly not important to the city's own position may well be crucial in looking at technology innovation and the broad smart cities scenario or the strategic decisions currently being adopted by international cities. To analyse a digital strategy for a particular city, the reader can draw on the book in a number of ways. First, the business models framework of Part I can be utilized. Second, the chapter or chapters from Part II that bear on the key dimensions of the city or sector can be used to provide some more specific guidance for business models design in data infrastructure. Finally, if smart cities designers are considering a decision, the reader can refer to the appropriate chapter in Parts III and IV. Even if a decision is not imminent, Part III will usually be helpful in reviewing decisions that have already been made and Part IV in examining the past and present decisions of other data infrastructure providers.

Throughout reading the book the reader will realise that a comprehensive analysis of data infrastructure for smart cities and its components requires a great deal of data, some of it will prove to be difficult to obtain. The book aims to provide the reader with a framework for deciding what data is particularly crucial, and how it can be analysed and used to design a suitable data infrastructure. Reflecting the practical problems of doing such an analysis which also involves many technical attributes; however, Section 10.3 (Complementary tools and techniques) provides an organized approach to actually conducting a smart city's study, including sources of field and published data as well as guidance in field interviewing.

This book is written for practitioners, which are leaders and decision makers exercising political or managerial leadership and control who could provide strategic leadership for a city or city partner, Chief Innovation, Information, Digital, and Technology Officers; academics in the field of smart cities, information systems and management of technology and innovation; consultants, technologists and entrepreneurs; and any government officials or observers who are seeking to understand digital strategy, data and technology in order to formulate public and digital policies.

1.1   Data infrastructure and strategy for smart cities

Essentially, designing a data infrastructure is developing a broad formula and strategy for how a city is going to provide city services, what its goals should be, who are the partner members of the value network and what governance policies will be needed to carry out these goals. To serve as a starting point for the reader before diving into smart cities and the data infrastructure's dynamic framework introduced by this

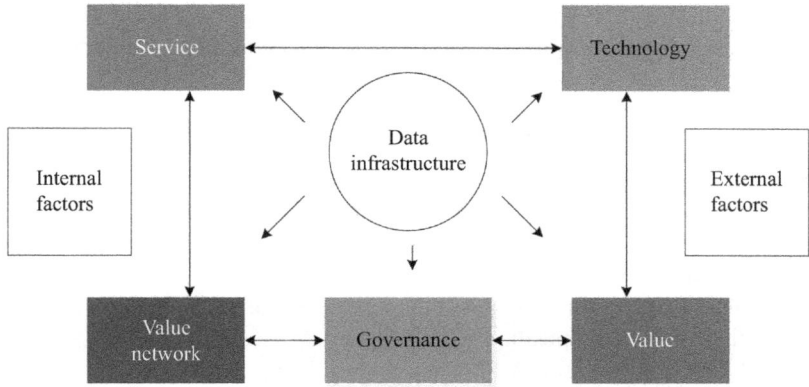

Figure 1.1   Context in which data infrastructures are designed

book, this section will give a very short introduction to the approach to design data infrastructures that will be presented in this book.

Figure 1.1 illustrates that broadly, designing data infrastructures involves the consideration of five key domains that determine the smart cities outcomes that a city can accomplish through the implemented infrastructure.

Service combined with technology determines the city's limits to the data infrastructure and strategy a city can successfully adopt. It represents its portfolio of value propositions and technology assets which will enable the exploitation of rich, interoperable and engaging cross-domain city data. The value represents the impact of the data infrastructure in delivering societal, monetary and environmental value to the society. It impacts government policy, social concerns and many others. The value network represents the collaborators and partners who will provide the expertise needed to deliver a data infrastructure, including standards, commercial exploitation, technological posture, data ethics and regulations and so forth. Governance determines opportunities and risks and defines the competitive environment in which the value network operates. These five domains must be considered before a city can develop a data infrastructure that is aligned with its long-term smart cities vision.

The design and delivery of a data infrastructure and strategy takes place in a dynamic process rather than static, and it is important to understand how they evolve over time. Its comprehension helps a city to understand the evolution of the competitive landscape in the face of changes in any of the five domains and the consequences that such changes may bring to the design of the data infrastructure and its respective outcomes. In our approach, data infrastructures design is divided into four phases:

1. **Research and development**: Conceptualisation and early prototyping.
2. **Procurement**: Strong engagement with the market and convert city's vision and digital strategies into business cases.
3. **Roll-out**: Small-scale market penetration and assessment.
4. **Market**: Large-scale market dissemination encompassing stages of market offering, maturity and decline.

An effective data infrastructure design and strategy can be determined by examining the value proposition and the existing processes, assets and policies.

Service
- Is the value proposition, which will be delivered through electronic channels, recognized as being better, and as outperforming competitions?
- Have the context of use and the market segment of users been identified?
- Do the services meet the value expectation of users (e.g. usability, feeling of security and accessibility)?
- Do the current relevant data regulations, policies and legal agreements address the goals and value proposition?
- Does the value proposition exploits commercial opportunities?
- What important governmental, social and political factors will present opportunities for business cases?

Technology
- Do the goals and value proposition match the technological and data resources available to the city?
- Is there a clear understanding on the available data, including its sources, volume, variety and temporal factors?
- Are there existing standards principles to which the providers of technology and data are committed to? Do they conflict with each other?
- Do the current relevant data regulations, policies and legal agreements address the goals and value proposition?
- Do the data management strategies comply with National, European and International (where possible) data protection regulations?
- Do the data management strategies comply with National, European and International (where possible) data protection regulations?
- Do the technical architecture relies on stable and well-defined interfaces to ensure interoperability between the platform, services and the applications provided by service complementors?

Value network
- Is the value proposition well understood by the members of the value network?
- Is there enough alignment between the value proposition and the values of the member of the value network to ensure commitment?
- Is it clear to the members of the value network the power they can exercise and how they will benefit from participating and collaborating in the data infrastructure development?
- Are the technical and architecture blueprint details available and understandable by the members of the value network?
- Does the timing of the goals and policies reflect the ability of the members of the value network to create actions?

Value
- Is the value proposition responsive to broader societal concerns?
- Do the business cases prioritise sets of data that can be used as trial of services, new standards and technologies and business models?
- Is cost associated with data collection known so that effective ways to recover costs of opening up data can be explored?
- Do the existing and future infrastructure support diverse business models (e.g. free, premium, subscriptions)?
- Is the technical and architecture blueprint based on open standards and open source technologies?

Governance
- Is there sufficient technical and managerial capability to allow for effective implementation?
- Are there processes and usage policies for ICT infrastructure so that assets reuse by the value-network members can be maximized?
- Do the governance and accountability arrangements ensure active participation of members, management of risks and value sources?

Although the list of questions presented above may be considered extensive and time-consuming; however, these questions make possible for cities to align the capabilities of their data infrastructures to their smart cities vision. As such, cities and stakeholders collaboratively apply their resources, expertise and insights to deliver services that offer unpreceded opportunities to solve social and environmental challenges. It is answering these questions that is the purpose of this book.

Part I

Smart cities and data infrastructures

Chapter 2
The evolution of urban intelligence

Studies on the future of technologies applied in urban spaces [5–7] along with the notion of global city wired society [8] and network society [9] have triggered the creation of a new field of activity which comprised professionals working at the intersection of *"people, place and technology"* [10]. Digital technologies offer opportunities to help cities achieve sustainable development [11]. This new field of activity has examined issues such as cultural, spatial and socio-economic repercussions and relationship of information and communication technologies (ICTs) and cities, local innovation systems and electronic applications (e.g. e-government, GovTech start-ups and incubators) [12,13], and the movement of digital to cyber to wired to intelligent to smart cities. The following section provides a historical background on the role that technology innovation has played in re-shaping cities and economies, and how it impacted the development of cities using the knowledge framework.

2.1 Context

Scientific research and technological innovation are acknowledged to be major driving forces of transformations in our societies [14], giving rise to economies, societies, cities, regions and countries. Cities provide an environment in which information flows rapidly and easily, making them a suitable stage for these transformations and primary disseminators and consumers of innovation [14–16].

Historically, the first identifiable cities and urban growth took place more than 3500 years BC in Mesopotamia [17]. As the precursors of urbanization, those urban populations already possessed sophisticated forms of political and social organization, literacy, and engaged with innovation progress predominantly in metals. To enable continuous transformations/developments and interaction between people and places, it was necessary to maximize the use of urban space and its resources. It was only during the Industrial Revolution in the eighteenth century that the modern urbanization process started in urban settings. Technology and science co-developed one another in many ways and they co-evolved while strengthening one another. The interaction between technology and science was made possible through the creation and dissemination of knowledge. Therefore, technology innovation assumed a predominant role in the generation of growth during the period of the Industrial Revolution and propelled the development of industrial areas surrounding suppliers of raw material and energy.

The last decades of the twentieth century represented a major shift in the global development and economic growth. Cities around the globe have experienced large-scale transformations in their economy, environment and society due to rapid globalization and urbanization. The world's population living in cities had increased to 50% by 2009 and is expected to reach 70% by 2050. Glaeser [18] argues that cities make people richer, smarter, greener, healthier and happier. In pursuit of these benefits, every day nearly 180,000 people are moving into cities and creating more than 60 million new urban dwellers every year. The current unbridled urbanization has transformed the optimization of urban spaces and resources, as well as the inter-action between people and places, into a critical global challenge. Cities of today with high density of population will need to be organized to enable a sustainable economic growth and guarantee a certain prosperity.

Alongside the rise in urban population, the increasing consumption has amplified problems associated with the scarcity of environmental and natural resources. As a result, cities have become very powerful drivers of environmental problems at local, regional and global scales. For instance, the built urban environment is responsible for 68%–80% of all energy consumption and greenhouse-gas emissions in the world [19]. Renewable and green sources of energy (biomass, hydro, wind, solar (thermal photovoltaic), geothermal and marine) are still not fully exploited, representing only 6% of total energy supply. Nuclear energy sources represent 8% of energy supply. Fossil fuels still represent over 86% of total energy supplies in the world today, and oil is at higher risk of becoming short in supply [20]. From 2030, there may be a supply crisis, requiring an increase of at least 30% in alternative sources of energy. Cities have a significant role to play in meeting global policy objectives and are challenged to maintain rising living standards for a growing population while significantly reducing the emissions we produce today.

In conjunction with the rise in urban population and the environmental changes, inadequate, deteriorating and ageing infrastructures are affecting the health of cities. The infrastructure of cities has evolved through many generations of technology that developed independently. The absence of connections between its component systems, which depend upon each other, often makes city utilities and services operate sub-optimally, limiting the creation of new value-added services, increasing transport costs and damaging existing logistics chains and economic models [21,22].

Van den Bessalar and Beckers [23] suggest that the success and failure of cities depend upon the technological infrastructure adopted by cities' policymakers. This dependence is said to be directly proportional to the size of the city: the larger the city, the stronger the dependence. Often, problems in city's infrastructure have fur-ther aggravated the many challenges associated with urban living in terms of law and order, security, health, waste disposal, housing, utilities, education, transportation and delivery of basic public services [24–28]. Some examples of cities that have failed due to deficient infrastructures are, for instance, New Orleans (flooding) and Brisbane (water management) [29]. Among the various challenges that city plan-ners, businesses and governments will have to be concerned about are the provision of more sustainable, secure and affordable energy; better infrastructure to manage energy supply and peak demand; meeting CO_2 reduction targets, more secure and

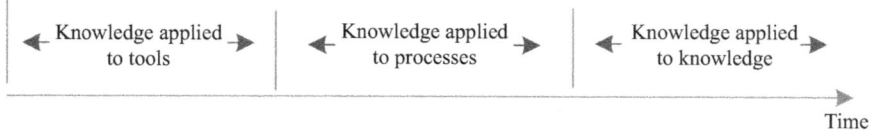

Figure 2.1   The knowledge framework [30]

high-quality water, more integrated public transport as well as the optimization of the existing assets.

2.1.1   Knowledge and technology innovation in cities

The creation, management and dissemination of knowledge and technology innovation have played an important role in the evolution of urban intelligence and may be the engine that will foster cities' development in today's world. Drucker [30] was one of the first scholars who saw the real significance of knowledge as becoming perhaps the most important competitive asset of modern enterprises during the 1950s. Drucker suggests that all the industrial revolutions represent a change in the meaning of knowledge. Technology and science influenced one another in many ways and they coevolved while strengthening one another. The interaction between technology and science was made possible through the creation and dissemination of knowledge.

At the first stage, knowledge was just applied to the development of tools, and as there was not too much scientific understanding at that time, it was not possible to understand why tools did not work nor how to improve them. The second stage involves the application of knowledge to improve the processes of how things are developed. At this stage, there is more understanding on how tools work and therefore they could be improved and used more efficiently. The latter stage is the stage where the developments made have scientific base, and therefore, innovative tools and more optimized processes could be realized. This represents the knowledge applied to previous knowledge and this is the stage where fast and more advanced innovations were developed and marked the start of the Digital Revolution. Figure 2.1 illustrates the knowledge framework used in this work which has been based on Drucker's definitions.

The evolution of urban intelligence after the Digital Revolution has been characterized by dramatic progress and developments in ICT and science and can itself be divided into four different periods. The first period corresponds to the emergence of the *digital cities*, a city where knowledge was applied to develop the virtual representation of cities and their endowment of hard infrastructure. At that time, too little was known about the repercussions and relationship of ICT and cities, and knowledge was principally applied to the development of tools. City policymakers invested heavily in a fragmented and unsatisfactory physical infrastructure and in the wrong business strategies: a top-down approach takes no regard for the human and social dimerism of cities.

With further advancement in ICT and the emerging idea of smart spaces, cities started to envision the provision of personalized services to users through the incorporation of ubiquitous computing and smart objects in the city physical infrastructure. This era gave rise to the term *ubiquitous cities* and was characterized by enhancements to the physical capability of the cities, which now could be accessible anywhere and anytime. A cyber-physical world started to emerge in which pervasively interconnected objects, things and processes had the potential to unlock unprecedented opportunities to optimally and transparently manage and operate the city infrastructures. There was not, however, sufficient knowledge on how to integrate such technologies into old and fragmented physical infrastructure of existing cities. Subsequently, a new movement which recognized the importance of the social and human capability of the cities and investment on innovation emerged. Hence, a third era has emerged with the creation of *intelligent cities* that support learning, collective and collaborative knowledge creation and innovation systems. This period is marked by the application of knowledge to improve processes by using and combining modern digital technologies with software, the city physical infrastructure and the community.

While studies on digital, ubiquitous and intelligent cities have focused on the physical capability (hard infrastructure) and/or on human, social and innovation capability (knowledge and creativity), recent studies of urban and technology evolution have moved cities towards a fourth period that envisions the use of modern digital technologies to fuel sustainable economic development. There is a growing demand for a more sustainable, efficient and liveable model in urban development [31,32]. In this period, a new concept named *smart cities* emerged. Smart cities aim at interconnecting cities' fragmented information systems and applications so that information can be integrated and exchanged across multiple processes, domains, systems and stakeholders that support better decisions, making control and management of resources, prompt reacting to unpredicted emergency situations and creating a more sustainable environment. The integration of city systems at the system-of-system level has been demonstrated to be able to create drivers for infrastructure innovation and improve the control of resources [22]. City infrastructure and systems that were previously invisible can then be discovered and improved through real-time information.

2.2   The evolution of urban intelligence

Economic historians suggest the economic growth that took place until the mid-eighteenth century was mainly due to institutional change and commercial progress rather than technology [33]. Although there was no change in technology at that time, economies managed to grow while at the same time maintained peace, law and order, improved communications and trust, among others [34]. The prime regulator of buildings height was the number of stairs a person might reasonably climb, and churches represented the tallest urban infrastructure on the urban environment. The noise in the streets came mostly from people, animals, and the wheels of carriages. Until this time, people have mainly relied on waterpower, using rivers and streams as sites for operations and favoured water for heavy transport. The economic growth and the shape of cities were determined based on trade and on the improved distribution of labour and land, which was enabled by institutional changes [35–37]. Until that time, the developments made had little or no scientific base, in which things were known to work rather than why they worked. As such, too little was known to create economic growth through technological innovation.

During the late eighteenth century in England the production process started to move from being craft-based to technology-based. People learned how to design better machines and applied new sources of energy to the industry, and new innovations gave rise to modern transportation and communication facilities, such as the railroad, telegraph, steamship and cable systems [33]. Factories could locate almost everywhere, as long as they had access to a railroad to deliver coal (for making steam) and to carry away finished goods. Factories also needed labour and a considerable pool of people started moving to industrial areas which represented a vast demographic change created by industry. This period was known as the *First Industrial Revolution*.

Further to these development, in the nineteenth century, a secondary wave of inventions emerged and novel ideas were applied into new and more industries and sectors. Innovations, such as the use of internal-combustion engine and electrical energy in the industry, resulted in decreasing the costs of production and transport, giving rise to the mass production and distribution systems of the *Second Industrial Revolution*. The late period of the second Industrial Revolution has been considered as one of the most fruitful and dense innovations in history [38, p. 22].

The Second Industrial Revolution extended the innovations of the first to a much broader range of activities and products, and as productivity climbed steadily, the price and cost of production diminished dramatically. Living standards, productivity, and the purchasing power of money increased rapidly, as the middle and working classes adoption of new technologies reaches like never witnessed before. In the early twentieth century, many technological advances were already in place such as aviation, telephony, electricity, synthetics and even electronics. Those technologies received continued development and improvement and were later transformed into useful tools in daily life. As populations grew dramatically and new businesses crowded in, cities began to grow both horizontally and vertically, giving rise to many skyscrapers in American cities. In the early twentieth century, the 1920 census showed that more Americans lived in cities that in farms, and between 1870 and 1920 the population of New York grew more than six-fold and Chicago's nine-fold, and at that time 13% of the world's population lived in cities. New means of travel also arose, first cars, streetcars and then subways, and railroads and transit lines encouraged the development of the new era of decentralization, and people looking for a better quality of life moved to greener, less polluted, quieter neighbourhoods.

The technological advancements persisted during the following century marking the start of the Third Industrial Revolution. Despite the two devastating world wars, an international economic collapse, violent inflation and a great depression, dramatic new developments occurred thanks to the microelectronics revolution which transformed mechanical and electro-mechanical systems into electronic systems. The real significance of this revolution became visible later when the companies learned to produce integrated circuits printed on silicon chips. The next big innovation was microprocessors, size components which were able to perform the functions of an entire computer. Advanced industrial countries went through a long-lasting period of

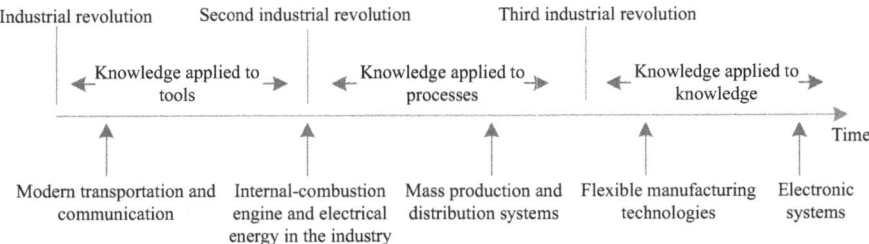

Figure 2.2   The knowledge framework: 1st to 3rd Industrial Revolution [30]

strong economic growth, high levels of employment and increasing affluence of their populations. This was also a time of rapidly expanding education, the building of new red-brick universities and the initiation of large-scale technology programmes. In the early 1970s microelectronics were first applied for military purposes, and shortly became a basis for the development of the so-called flexible manufacturing technologies. Another important new area of commercial applications was the development of personal computers.

The main technology innovations which took place during the three Industrial revolutions are summarized in Figure 2.2. Technology innovation assumed a dominant role in the generation of growth during the period of the Industrial Revolution. According to Drucker [30], the first industrial revolution is the application of knowledge to tools, processes and products (e.g. transport, communication). The Second Industrial Revolution is the application of knowledge to work and processes resulting in the production revolution. The last one is the application of knowledge to the knowledge itself – derive new knowledge from previous knowledge, which was the time where modern manufacturing technologies and electronic systems were created and applied across many industries.

2.2.1   The digital revolution

During the 1980s, many innovation researchers directed their attention towards the digital revolution, which was regarded as a major technological innovation that followed the microelectronics revolution [39]. The microelectronics revolution began during the late 1940s when the first transistors were built at the Bell laboratories in California. The essence of the digital revolution is in the generalization of a new way of processing information – based on binary logic – and in the development of increasingly effective ways of handling and delivering such information. This process led to the construction of "*information highways*" and increased the synergies between previously isolated and fragmented technologies (e.g. computers, electronic media, etc.). At that time the digital revolution was perceived to carry a great potential to generate new kinds of needs and an entirely new demand for integrated technologies. The innovation potential of this new technological wave was not only in bringing together existing technologies, but rather to give rise to the development of new software, services, and contents that are being brought together. The way in which this new information and technologies were used had the potential to create competitive advantage among economies, cities and countries [40].

With the advancements of modern digital technologies, cities have changed the way in which they interact with their inhabitants, businesses and service providers by the exchanging of a vast amount of data every day. The historical period of ICT development was characterized by the utmost speed at which the new technologies of communication and information have been diffused, representing a development never seen in a very short period of time approximately two decades. The digital revolution and technology innovation alongside the efforts to benefit from technology in urban design and planning gave rise to a new form of urban infrastructure.

In the broad area of city sciences, there are many definitions of digital, ubiquitous, intelligent and smart cities. As a result, there is a real difficulty in understanding the meaning of those initiatives. Differences, sometimes substantial, emerge from the fact that stakeholders, businesses and policymaker's backgrounds and their approaches to the issue under the perspective of a capital, depending on their specific interests and expectations. The Oxford dictionary defines capital as *"A valuable resource of a particular kind"*, and The Free Dictionary defines it as *"assets available for use in the production of further assets"*.

Through the evolution of urban intelligence, we note that the development of a capital – referred in this book as **capabilities** – enhanced and/or produced new resources which are the essential foundations of the cities of the future. The seven capabilities discussed in this book are:

Social capability: Halpern [41] defines social capability as being composed of *"a network; a cluster of norms, rules, values and expectations; and sanctions"*. Putnam [42] argues that stocks of social capability tend to be self-reinforcing and cumulative, i.e. those who have social capability tend to accumulate more. Social capability is also what the social philosopher Albert O. Hirschman calls a *"moral resource"*, that is a resource whose supply increases rather than decreases through the use and, unlike a physical resource, does not become depleted if not used.

Physical capability: The physical resources are consisted by the city's endowment of hard infrastructure. Van den Bessalar and Beckers [23] suggest that the success and decline of cities depend upon the technological infrastructure adopted by city policymakers. This dependence is said to be directly proportional to the size of the city: the larger the city, the stronger the dependence. The physical capability can exacerbate inequality due to resource allocation, investment and maintenance being partial to perceived sources of value creation and competitive articulation in urban economies [12,43]. Modern physical resources (e.g. digital technologies) offer a new wave of opportunities to mitigate some of these impacts and create a balance between social, environmental and economic opportunities that will be delivered through smart city planning, design and construction.

Human capability: Relates to the intelligence, inventiveness and creativity of the individuals who live and work in the city. This perspective was described by

Richard Florida [44] as part of the *"creative city"* 0, which gathers the values and desires of the new creative class made by knowledge and talented people, scientists, artists, entrepreneurs, venture capitalists and other creative people, which have an enormous impact on determining how the workplace is organized, whether companies will prosper, whether cities thrive or wither.

Innovation capability: The innovation capability builds on the conception of innovation systems created by Lundvall [45,46] that identify cities as the main trigger of economic growth in a globalised world achieved through innovation. Innovation resources require highly coordinated information exchange, network creation among actors (e.g. government agencies, businesses, funding bodies) and is a continuous learning process as the strategic value of information changes in time; thus, a regular knowledge update is important [47].

Institutional capability: The institutional capability comprises the institutional mechanism of a city that supports businesses and citizens to achieve optimum intellectual performance, life, mobility and innovation. The institutional resources facilitate cities to form, develop, and use innovation and social resources in a coordinated way. Institution resources rely upon the support of governance and policy. Cooperation between stakeholders and institutions is very important to develop and to intervene efficiently to meet challenges posed by a globalized economy and society. According to Morgan [48], active cooperation can shift the emphasis from the power to decide to the power to transform (i.e., deliver), key to overcoming the delivery deficit.

Economic capability: Cities are seeking for intelligent solutions that will provide a high quality of life without excessive operating costs and enable them to evolve and appropriately react to environmental changes and recover from unpredictable situations while meeting ambitious targets for innovation and sustainability agendas. The economic capability constitutes the money and assets received from cost-effective government, improvement of services, creation of new and innovative services, stimulation of competitiveness, stimulation of a more active participation and empowerment of citizens, and the re-use of information across multiple processes, systems, value chains and stakeholders.

Sustainable capability: Environmental issues are near the top of all cities agendas [49–51]. As the quality of life becomes an important source of competitive advantage, cities must provide a clean, green and safe environment for their citizens. Planning, transport, finance and economic policies all need to reflect the environmental goals that a city sets for itself. On the other hand, citizens also need to be engaged in the development and implementation of environmental policies and be encouraged to take responsibility for the quality of the environment in which they live.

In this book, we highlight the evolution of urban intelligence through technology innovation, and how seven urban capabilities have been developed through the many stages of the evolution of urban intelligence.

Table 2.1   Digital cities definitions

Definition	Reference
"A community digital space, which is used to facilitate and augment the activities and functions taking place within the physical space of the city"	[52]
"A digital city is substantively an open, complex and adaptive system based on computer network and urban information resources, which forms a virtual digital space for a city. It creates an information service marketplace and information resource deployment center"	[53]
"A connected community that combines broadband communications infrastructure; a flexible, service-oriented computing infrastructure based on open industry standards; and, innovative services to meet the needs of governments and their employees, citizens and businesses"	[54]
"An arena in which people in regional communities can interact and share knowledge, experiences, and mutual interests. Digital City integrates urban information (both achievable and real time) and create public spaces in the Internet for people living/visiting the city"	[55]
"A network of organizations, social teams and enterprises operating within an urban area"	[55]

2.3   Digital cities

The digital era of cities has begun in the early 1990s with the emergence of several community networks in the US and Canada. In the US particularly, the free-net initiative was the first step towards the diffusion of the electronic provision of information within American cities. The term digital cities were, however, first used in 1994 in the Amsterdam case and it has conceptualized the digitization of cities in Europe [23]. There are several definitions for the digital cities in the literature, and we highlight the ones presented in Table 2.1.

Other scholars consider the digital city as the transformation of cities and urban areas using modern technologies to provide novel and interactive city-wide services, which provide information and transactions to governments, citizens and business [56,57].

We define Digital Cities as
"A digital environment holding information gathered from a small community and interactively released through an Internet portal".

Our definition is aligned with the views from other authors in urban sciences [58–60]. The main reason for the development of digital cities worldwide has been to improve the way in which cities interact with their inhabitants, businesses and service providers. Digital cities were implemented through new and online organizational and economic structures within pre-existing social and political structures [61,62] and were established under the umbrella of e-governance [63,64].

One of the earliest digital cities developments took place in Kyoto through the development of modern digital systems to serve as a social and information infrastructure for urban everyday life [65]. Other examples of digital cities initiatives include Seattle [66], Shanghai [67], Virtual Helsinki [68] and Amsterdam [23].

This variety of developments is related to the different social contexts in which digital cities have developed [23]. For instance, one of the earliest developments took place in Kyoto as a project sponsored by NTT with the aim to create next-generation systems for digital communities that could serve as a social and information infrastructure for urban everyday life. The Digital City Amsterdam was created to serve as a virtual space representing the real urban setting to provide a space for public communication to its dwellers. The Digital Helsinki focused on developing a modern metropolitan network. Some examples of digital cities developments include virtual city representations and e-government services which support various public policy goals such as information on traffic, parking, social inclusion or environmental awareness [56], and information city representations in which information is obtained from communities and offered to the public by online portals [58,59].

Other digital cities initiatives included the direct participation of community members in the design process of urban planning. The Global Digital City Network (GDCN) is an international organization between *knowledge-based* digital industrial cities. It has been designed to encourage the sharing and exchange of information and technologies in the multimedia (and other knowledge based) sectors to promote the competitiveness of its member cities. Current members of GDCN, in addition to the founding Digital Dundee City (UK), are the Digital Cities of: Chunchon (S. Korea), Kakamigahara (Japan), Gold Coast (Australia), Nizhny Novgorod (Russia), Jilin (China), Shenyang (China), Yanji (China) and Taipei (Taiwan). A similar international network of digital communities has formed the International Network of e-Communities group (INEC). According to Boschma [69],

> "[…] it is meaningful to talk about regional competitiveness when the region affects the performances of local firms to a considerable degree. This is especially true when the competitiveness of a region depends on intangible, non-tradable assets based on a knowledge and competence base embedded in a particular institutional setting that are reproduced and modified through the actions and repeated interactions of actors".

Some researchers have focused on digital cities ICT adoption [70] and proposed decision support methodologies to assist digital cities in adopting ICT best practices [71]. A good account of the architectural design of digital cities is described in the work of [56]. The author presents a three-layer model to design a multi-purpose digital city. The first layer is called the information layer, which is concerned with the integration of data from different data providers (e.g. real-time sensory data, media, geographical data, etc.). The second layer is the interface layer and contains maps,

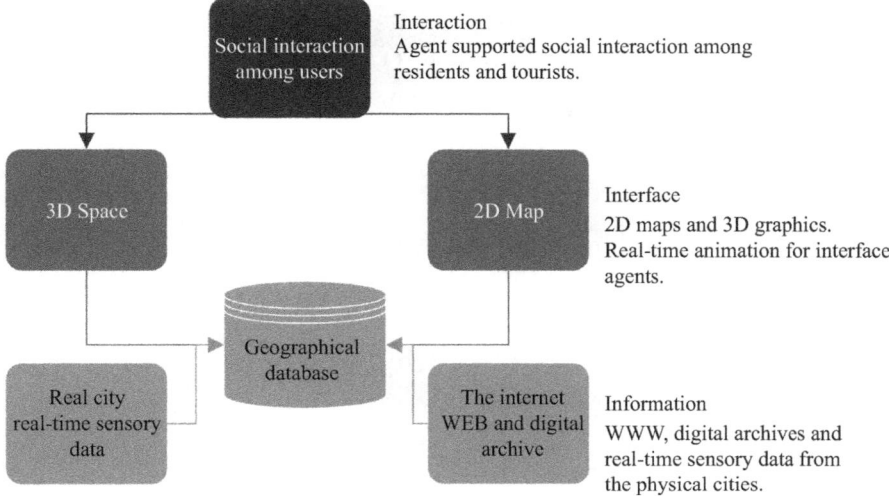

Figure 2.3   The architecture of digital cities [56]

3D virtual spaces, car, buses, trains, avatars representing citizens, among others. The third layer is the interaction layer where there is the interaction and information exchange between citizens and visitors. Figure 2.3 illustrates the system architecture of digital cities adapted from Ishida [56].

The architectural design of Digital cities is not uniform and varies according to the digital services that a city is willing to offer. For instance, the universal architecture illustrated in Figure 2.3 does not reflect the genuineness of a real city. Yet it shows an architectural design in which the digital city is fully controlled by a central organization (e.g. city council, businesses). In a real city, this pattern is not observed as many third-parties are involved in the creation and provision of information flows within modern cities. For other cases of digital cities (commercial, policy-driven and virtual), the architectural design is simpler and reduced to a directory of urban information, communication platform, web forum or a collection of visual data. When a physical city is transferred to the digital space it cannot be just represented by a web portal.

In contrast, according to Komninos [52], the union of various websites collaborating to the form, activities and roles of a city can be considered as a digital city. In order to overcome the constraints of this non-uniform architectural design, Komninos [52] devised a four-layer general model of digital city. The author claims that all combinations and alternative architectural design of a digital city may derive from this universal model as illustrated in Figure 2.4. The four layers' universal model of digital city is defined as follows: The first layer is the information storehouse, which represents a database that encompasses all kind of digital content (e.g. text, images, diagrams, video, and multimedia). The organization of this content is based on logical patterns, the districts and the city hierarchy.

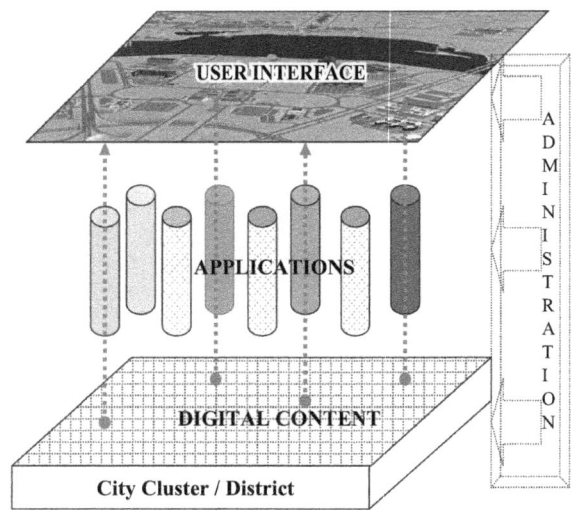

Figure 2.4   An extended architecture for digital cities [52]

Other researchers have proposed alternative solutions to facilitate information dissemination among government, citizens, and other stakeholders [72–74], while others have focused on technical platforms (mainly Service Oriented Architecture (SOA)) [75–77], ontologies for data modelling [74], and parallel computing to increase applications performance [78]. More recently, some researchers have focused on digital cities ICT adoption [70] and proposed decision support methodologies to assist digital cities in adopting ICT best practices [71].

2.3.1   Capabilities development in digital cities

The concept of digital city seems to integrate and/or evolve the following capabilities:

1. **Physical capability:** Digital cities have mainly focused towards technological adoption and the development of new Internet applications to support local governments, and have tried to use the physical endowment of cities to impact society, urban planning and city competitiveness.
2. **Institutional capability:** We can argue that digital cities have shifted the institutional capability of cities by accelerating governments to adopt e-Government Processes in order to improve their governmental operations and transactions.

2.3.2   Structural analysis of digital cities

While people and citizens (human capability) are cited in some digital cities initiatives, their role is less proactive and seen primarily as consumers. The available definitions for Digital Cities do not seem to be focused on enhancing the quality of life of citizens. We argue that this illustrates how the basic premise of integrating ICTs with city development views citizens, primarily as consumers. In this context, the main research on digital cities focuses on technological infrastructure and Internet applications, and on the evaluation of their effectiveness in addressing the determined policy goals [79], considering that ICT will somehow impact on social and cultural urban practices rather than vice versa.

Infusing modern technology into urban settings without assessing its effects on the various urban capabilities is not the best tactic to follow. By failing to follow appropriate business strategies, many cities have reduced the likelihood that they could succeed. Take the case of one of the first digital cities initiatives, Blacksburg Electronic Village, in which a consortium created by Virginia Polytechnic Institute, State University, Bell Atlantic, and local authorities aimed to create a virtual information space close to the region. They have spent the first two years preparing computers and communication equipment, and dial-up services started in 1993. Naturally, it has attracted the attention of many people; nonetheless, in this case, the leading role was taken by university, organizations and administration as the network was constructed from their technological perspective. So, non-techie citizens could not take the lead. Two years later, the activity of Blacksburg Electronic Village decreased because of the disagreement between stakeholders (technology providers and users) in regarding the goals and expectations of community networks [55]. The South American city of Lima has increased rates of telecommunication diffusion enormously to rise technology take up, however, in 1990 only 7% of the households had access to the Internet, and deprived citizens were 50 times less likely to have access to the Internet [80]. The technology diffusion does not necessarily mean acceptance nor equal take up.

Digital city represented the application of knowledge to create the physical capability of cities in a fragmented fashion, as the application of technologies (e.g. remote sensing, GIS, wireless communication) was not fully associated between them nor integrated within the urban infrastructure planning. Hence, such applications have not met the major requirements to build an integrated physical and virtual environment [52]. Digital cities are often reduced to a directory of urban information, communication platform, web forum or a collection of visual data; however, these approaches do not reflect the genuineness of a real city but define digital cities as a fully controlled by a central organization (e.g. city council, businesses). When a physical city is transferred to the digital space, it cannot be just represented by a web portal. In a real city, this pattern is not observed as many third-parties are involved in the creation and provision of information flows within modern cities.

2.4   Ubiquitous cities

ICTs have played an increasingly important role in the planning, management and optimization of urban infrastructure. In the 2000s, the technology development in those areas as a result of tremendous changes in mobile networks (i.e. mobile phone, vehicle navigation, smart card and personal tracking system) [81]. In particular, mobile phones became intelligent devices, which are used for not only interpersonal communication but also for accessing information and services provided by the Internet.

Table 2.2   Ubiquitous cities definitions

Definition	Reference
U-city is to create a built environment where any citizen can get any services anywhere and anytime through any ICT devices	[87]
A ubiquitous city is defined as "a next generation urban space that includes a integrated set of ubiquitous services: a convergent form of both physical and online spaces". These services ultimately aim to enhance quality of life factors, such as convenience, safety and welfare	[91]
"U-City is a 21st century futurist city which enables the service such as one stop administration service, automatic traffic, crime prevention, fire prevention system and home-networking of residential places which fused high tech infrastructure and ubiquitous information service into the urban area"	[92]

Advancements in wireless technologies alongside with the embedded physical devices and software applications in everyday settings and their linkage with common-place tasks have brought the emergence of ubiquitous/pervasive computing, intelligent environments and smart spaces [54]. These advanced technologies provided opportunities for the people, so they can communicate not only with other people but also with any product or services of the existing urban infrastructure (i.e. transport, waters, public parks, hotels and path directions) if they contain sensors, processors and software. These physical infrastructure and their objects can be self-monitored, controlled and protected by digital networks and can be communicated with citizens and other physical objects.

Ubiquitous and pervasive computing is a user-centred paradigm that has a main objective to provide the user, in the most natural way, all the time everywhere access to information and computer services [82–86]. Digital sensors and tags alongside mobile networks (i.e. mobile phone, sensors, smart tags and cards and vehicle navigation) have been widely used for purposes other than communication but also access information and services provided by the Internet [87]. These modern digital technologies provided opportunities for people to communicate with machines and services within the existing smart space. Smart space is one that can improve inhabitants experience in an environment by acquiring and applying knowledge in this environment [54]. The extension of smart spaces to a city-wide perspective has resulted in the creation of the ubiquitous city (u-city).

A u-city is a digital city in which urban services are delivered through ubiquitous sensing and communication resource capabilities, such as in residences, building infrastructure and open spaces [87]. It represents a futuristic city which integrates urban spaces with ubiquitous technologies to provide a high quality of services regardless of time and location [81,87–90]. Table 2.2 presents other u-city definitions in the literature.

Technologies that have enabled the development of u-cities include Ubiquitous Sensor Network (USN), Context Awareness Computing, Broadband Convergence

Network, Augmented Reality (AR), Geographic Information System and Wireless Broadband. South Korea and Japan have been the pioneers in the development of u-cities and have inspired other countries to adopt the concept of embedding ubiquitous computing on the urban physical infrastructure [93]. Korea has continuously developed national strategies for sustainable urban development through different projects such as the U-Korea initiative, one which aims at integrating the physical and digital layers of the city using a wide wireless urban infrastructure network [94].

South Korea and Japan have been the pioneers in the development of u-cities which focused on providing everyone with an opportunity to access urban infrastructure and services by using any information technology devices regardless of time and location. During the last two decades, Korea has continuously developed national strategies for sustainable urban development through different projects (i.e. Cyber Korea, E-Korea and U-Korea). U-Korea initiative focused on the integration of ICTs and the physical city to aim at a convergence of the virtual and actual cities through the use of a wide wireless urban infrastructure network [94]. Technologies that have enabled the development of u-cities include Ubiquitous Sensor Network (USN), Context Awareness Computing, Broadband Convergence Network, Augmented Reality (AR), Geographic Information System and Wireless Broadband.

Lee *et al.* [87] proposed the U-eco City Integrated Service Management Platform (ISMP-UC) for ubiquitous city (u-city), in which urban infrastructure is composed of people and objects connected through environmentally friendly technologies, providing services to an effective linking of people, environment and technological advancement. The Seoul's Digital Media City (DMC) is a ubiquitous street where individuals can display and have hands-on experience with innovative digital media appliances. In Singapore, the One-North project aimed at creating a more efficient power system. Other academic researchers have proposed solutions to integrate the urban engineering (e.g. road traffic, environment and GIS) with pervasive technologies [95–97], as well as investigating physical and technical countermeasures against the security threats, such as privilege, confidentiality, integrity, denial blockade, availability and information security [98].

We define ubiquitous cities as

"A city which is built digital from the ground up and is embedded with modern digital pervasive technologies that enables real-time delivery and personalization of services throughout the urban environment".

2.4.1 Capabilities development in ubiquitous cities

The concept of ubiquitous city seems to integrate and/or evolve the following capabilities:

1. **Physical capability:** Ubiquitous cities improved the physical capability of cities by further developing and researching new communication infrastructures, and the convergence of smart spaces with the city space.

2. **Human capability:** Ubiquitous cities have considered the human capability for enabling customized services for citizens at their convenience – the concept of anywhere any time connectivity – and matching user needs with the right ubiquitous services is crucial in ubiquitous cities [91,99].

2.4.2   Structural analysis of ubiquitous cities

Ubiquitous cities aim at creating complex information systems which can deliver services from anywhere to everyone. Likewise, digital cities, ubiquitous cities are heavily concerned with the delivery of high-speed networks and powerful underlying technologies the physical capability. Despite the enormous potential in creating such ubiquitous/pervasive cities, the challenges imposed by its conception in contemporary cities remain an open issue. Ubiquitous cities have been architected in the first decade of the twenty-first century as company towns, completely smart from the ground up [10]. Embedding new technological infrastructure in old cities and making it work in favour of its human and social capability is very challengeable, due to its heterogeneous, ageing and overlapping technological infrastructure making the ubiquitous computing system not to be addressed in an integrated manner by policymakers and urban planners [27,100]. Furthermore, there are several additional obstacles and challenges for u-infrastructure to become a seamless infrastructure provision and management system, including financing the technology, the digital divide, the knowledge divide, openness of the political system for such a transparent system.

Moreover, ubiquitous cities' initiatives are mainly top down, a technology-push similar to the digital cities movement, and the social, cultural and innovation capabilities of u-cities are still non-existent. This puts a significant barrier to secure an interaction between the technology and social systems, limiting people to unlock their collective knowledge to drive innovation and the creation of value. As pointed out by some authors [3,101,102], some cities have used technology to shift the basis of urban intelligence away from its essence. While there are evidences on the impact of ICTs on urban environment [12,103], there is a belief that information technology will on its own deliver a kind of technological *"Field of Dreams"* scenario.

For all these reasons, the concept of ubiquitous cities has not been widely adopted and only managed to be by very few countries. The disregard of social construction processes (social and innovation capability) that shapes technology usage by humans (human capability) [104] has given rise to more qualitative approaches which analyses the interplay of the social, human, innovation and institutional capabilities in local settings [23,105,106].

2.5   Intelligent cities

In 2005, Van den Besselaar *et al.* [23] stressed that digital cities would develop at the same pace as computer and network technologies and it certainly did. The development of new and modern ICT technologies has stimulated the digitization of modern cities and made available an enormous pool of information, transforming our society into what is known as *"information society"* or *"knowledge-based society"* [9]. From the crossing of the ubiquitous city with the knowledge society, a new kind of city known as intelligent city has emerged. Among the several definitions of intelligent cities, we highlight the definitions presented in Table 2.3.

An intelligent city depends upon a modern physical infrastructure and a community with crucial knowledge and innovation capabilities within organizations to survive in the actual competitive global economy. The main goal of intelligent cities initiatives is not merely to employ state-of-the-art technologies in a city physical infrastructure, but rather, it is focused on preparing the social community to meet the challenges of a global, knowledge economy [79]. The main distinction between a digital and an intelligent city lies in the ability to the latter in support creativity, experimentation, learning, technological development and innovation.

Komninos [52] suggests that an intelligent city is a multi-player territorial innovation system, and to illustrate this organizational structure, he proposes a three-layer architecture that interlinks innovative clusters with the digital community spaces, as it is illustrated in Figure 2.5. The first layer of this architecture relates to a city's productive clusters in manufacturing and services. The second layer is formed by the city's institutional mechanisms which regulate and coordinates knowledge flows, as well as the co-operation in learning and innovation. The final and third layer is composed of technological infrastructures and digital tools, and spaces for learning, experimentation and innovation.

Table 2.3 Ubiquitous cities definitions

Definition	Reference
"Territories with high capacity for learning and innovation, which is built-in the creativity of their population, their institutions of knowledge creation, and their digital infrastructure for communication and knowledge management"	[52]
"Has all the infrastructure and infrastructure of information technology, the latest technology in telecommunications, electronic and mechanical technology […] used to unite, promote, acquire and higher circulation of information and quality of living all together"	[107]
An intelligent city can be defined as a "city of knowledge where technological innovation and people's creativity are supported and encouraged, with strong institutional leadership and organizational capacity, creating the best possible conditions to increase competitiveness and sustainability"	[47]
"Intelligent cities offer skills, institutions and virtual spaces of cooperation sustaining the creation of new knowledge (research), monitoring knowledge flows (intelligence), disseminating existing knowledge (technology transfer), applying knowledge (innovation), developing new activities based on knowledge (incubation) and managing knowledge remotely (e-government)"	[108]

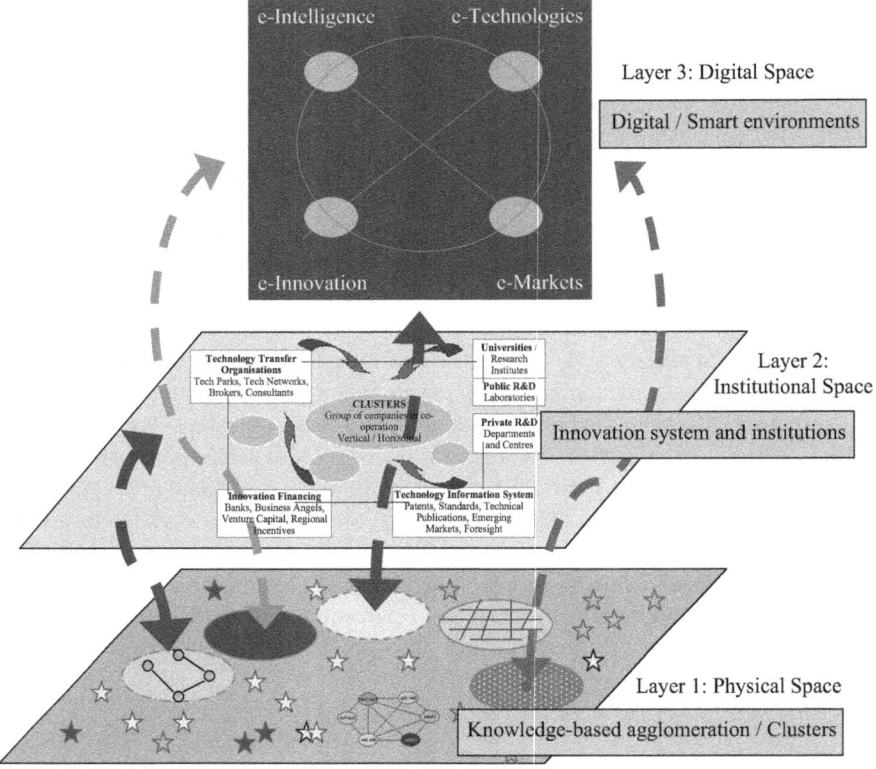

Figure 2.5   Intelligent cities as multi-player territorial innovation systems [52]

We define intelligent cities as

"A city which has digital components and seeks to strengthen the role of the citizen in the innovation cycle, facilitating technological and social innovation, converging on common quality and standards and, methodologies in open innovation geared to the citizen".

In the UK, the *myedingburgh.org* project provides a user-friendly information portal and community space for learning, which is referred to as "*CGfL*". In this portal, communities have access to information regarding their cities' planning, development and design, as well as opportunities to engage in local community decisions related to the city's urban regeneration strategy. This initiative has similarities with the concept of *Living Labs*, which aims to engage the local communities in the entire course of innovation and development of products and services in the city. A very good example of Living Labs is the *Open Living Labs* in Europe[1] which is a project dedicated to increasing the regional competitiveness of European cities on the global stage and at the same time help their communities to take greater control of their environment and sustainability goals [109]. As of 2017, there were over 340 Living Labs throughout Europe making use of modern digital technologies to connect communities with a wide range of developers and policymakers.

Several research consortia and academic institutions worldwide are currently working on developing platforms and solutions to deliver the promise of intelligent cities. Projects include the Crossroads Copenhagen (Denmark), Digital Media City (Korea), MIT Environs (USA) and Sapiens (Brazil).

2.5.1   Capabilities development in intelligent cities

Thus, the concept of intelligent city seems to integrate and evolve the following urban capabilities:

1. **Physical capability:** The mobile wireless and Metropolitan Area Networks (MANs) developed in digital/ubiquitous cities initiatives has served as the telecommunications backbone over which cities can further develop, enhance and research new communication infrastructures, digital pervasive spaces and information management tools. The many digital systems implemented in intelligent cities provides the means to coordinate the various silos of government and business services which sustain new urban functions to support knowledge creation, technology transfer and innovation.
2. **Human capability:** This capability was initially developed in some ubiquitous cities initiatives, and intelligent cities have shifted this capability by providing digital solutions capable of fostering intelligence, inventiveness and creativity of the cities' inhabitants.
3. **Social capability:** This capability was developed as a result of the embedded routines of social co-operation, knowledge creation and sharing. The development of human and physical capabilities also paved the way for the creation of Living Labs and other community-driven initiatives by introducing the notion of openness, realism and empowerment of users and the co-creation of new solutions [110]. These developments are facilitated by the institutions that are in place in the cities and are the fundamental drivers of cooperation in knowledge and innovation. The social capability emphasizes that the intelligence of cities are collective rather than an individual achievement [111].
4. **Institutional capability:** The organizational issues related to the management of innovation prevail over the technological ones of setting up virtual environments and technology systems [112]. Research on intelligent cities suggest that governments must connect the key stakeholders involved in public policy design, decision-making and delivery, as well as foster a two-way communication among citizens, socio-economic agents and other public institutions.

---

[1] http://www.openlivinglabs.eu/

5. **Innovation capability:** The innovation capability in intelligent cities resulted from the development of social and human capabilities, alongside smart spaces and technologies which developed further local creativities, knowledge creation and sharing, networking, experimentation and innovation. A city local system of innovation is enhanced by digital collaboration spaces, interactive and accessible tools,[2] and embedded systems that can manage the physical infrastructure.[3] The city gains innovation capability, which is translated into increased competitiveness and ecosystems of businesses, a more liveable and sustainable environment, more jobs and wealth [112].

2.5.2 Differentiation factors of intelligent cities

The urban intelligence seems to have evolved to the intelligent cities category because of the growing body of understanding on the convergence of the physical, innovation and social and human capabilities of cities. The development of innovation capability has motivated the formation of global scale knowledge networks between cities main stakeholders (e.g. government agencies, businesses, funding bodies) that is dependent on social and innovation capabilities, local institution settings, culture and ICT. Research in this field has aimed to strengthen the competitive position of cities and regions [113] enabling them to succeed in the global landscape. While studies on digital cities and intelligent cities have mainly focused on the physical capability (hard infrastructure) and/or on human, social, institutional and innovation capability (knowledge and creativity), recent studies on urban evolution along with the adoption of ICT has been moving cities towards using modern digital technologies to embed smartness into city infrastructure and assets to fuel sustainable and economic development. These initiatives have originated the concept of smart cities which tries to integrate and evolve the urban capabilities to accommodate the sustainable and economic capabilities.

While studies on digital cities-to-intelligent cities have mainly focused on the physical capability (hard infrastructure) and/or on human, social, institutional and innovation capability (knowledge and creativity), recent studies on urban evolution along with the adoption of ICT have been moving cities towards using modern digital technologies to embed smartness into city infrastructure and assets to fuel sustainable and economic development. These initiatives have originated the concept of smart cities which tries to integrate and evolve the urban capabilities in order to accommodate the sustainable and economic capabilities. The use of modern technological infrastructure to wisely manage natural resources while improving citizen's life and city infrastructure has brought the concept of smart cities [49,114]. Research on smart cities has focused on the socio-technical and socio-economic aspects of urban growth [115,116], the physical-digital infrastructure to improve services [22,52,117] as well as the mitigation of the effects of environmental change [117,118].

---

[2] http://www.cityzen-smartcity.eu/hello-world/, https://www.playthecity.nl/
[3] http://blindmaps.org/manifesto/

Chapter 3

Smart cities

The unprecedented speed in urbanization along with the current economic crisis and the ageing infrastructure of cities has stretched cities to the limit which are suffering to provide basic urban services. Cities of today with high density of population will have several challenges not experienced so far and will need to be organized in a strategic way to enable a sustainable economic growth and guarantee a certain prosperity. Scholars have outlined the growing demand for a more sustainable, efficient and liveable model in urban development [31,32]. The concept of smart cities has emerged due to increasing interest in researches on the innovative socio-technical and socio-economic aspects of urban development [115,116], the physical-digital infrastructure to improve the quality of urban services [22,52] and the mitigation of the effects of environmental change [116–118]. The literature on the smart city is fairly new and some publications have theorized and described the usage of this concept [3,52,119].

As of 2012, approximately 143 '*self-designated*' smart city projects were under development or under completion in the US, Europe and Asia [120]. The variety of views about what a smart city is has resulted in broad definitions [1,3] with no focus on specific technologies, strategies or cities. Some focus on digital technology as the main driver and enabler of smart cities, while broader definitions include socio-economic and the citizen's participation in sustainability agendas, quality of life and urban welfare. What most smart city definitions have in common, however, is that they consider the use of new technologies and data as the means to solve cities' economic, political, social and environmental challenges. Some definitions of smart cities are presented below and are divided into broad definitions, technology-, data- and citizen-driven definitions, as shown in Tables 3.1–3.4.

We define smart cities as

'The use of modern digital technologies to improve city services, businesses, city infrastructure, the life of citizens while wisely managing natural resources and supporting urban capabilities.'

Research on smart cities focus on identifying ways for city systems (e.g. energy, water, transport, environment, waste and recycling, health care) to work in an orchestrated manner so that cities are a better place to live, where business can prosper, citizens are happy and healthy, and there is a sustainable economy [1,52,130,131].

Table 3.1   Smart cities' definitions – broad definitions

Definition	Reference
A process in which social capability, increased citizen engagement, modern infrastructure, and digital technologies create more liveable and resilient cities which can respond quicker to new challenges	[121]
A Smart city is made by a combination of six characteristics: smart environment, smart governance, smart economy, smart living, smart people, and smart mobility	[4]
A 'well-defined geographical area, in which high technologies such as information and communication technology (ICT), logistic, energy production, and so on, cooperate to create benefits for citizens in terms of well-being, inclusion and participation, environmental quality, intelligent development; it is governed by a well-defined pool of subjects, able to state the rules and policy for the city government and development'	[122]
A 'city that invests in human and social capability and traditional (transport) and modern (ICT) communication infrastructure fuel sustainable economic growth and a high quality of life, with a wise management of natural resources, through participatory governance'	[1]

Table 3.2   Smart cities' definitions – technology driven

Definition	Reference
'Smart City is the product of Digital City combined with the Internet of Things'	[123]
'Scalable solutions that take advantage of information and communications technology to increase efficiencies, reduce costs, and enhance quality of life'	[124]
'The use of smart computing technologies to make the critical infrastructure components and services of a city which include city administration, education, healthcare, public safety, real estate, transportation, and utilities more intelligent, interconnected, and efficient'	[121]
'Smart Cities combine ICT and Web 2.0 technology with other organizational, design and planning efforts to dematerialize and speed up bureaucratic processes and help to identify new, innovative solutions to city management complexity, in order to improve sustainability and liveability'	[31]
'A city in which it can combine technologies as diverse as water recycling, advanced energy grids and mobile communications in order to reduce environmental impact and to offer its citizens better lives'	[a]

[a]Setis-Eu. (2012). setis.ec.europa.eu/implementation/technology-roadmap/

It is thus no surprise that smart city and its data have attracted the attention of many researchers [2,27,31,132–135], and large corporations such as IBM, SideWalk Labs (Alphabet), Cisco Systems, Accenture, Arup and Siemens, among others. For instance, IBM's approach to smart cities revolves around three main concepts: instrumentation, interconnection and intelligence [136]. The instrumentation principle

Table 3.3   Smart cities' definitions – data driven

Definition	Reference
'A city that monitors and integrates conditions of all of its critical infrastructures, including roads, bridges, tunnels, rails, subways, airports, seaports, communications, water, power, even major buildings, can better optimize its resources, plan its preventive maintenance activities, and monitor security aspects while maximizing services to its citizens'	[125]
'Smart city is one that makes optimal use of all the interconnected information available today to better understand and control its operations and optimize the use of limited resources'	[126]

Table 3.4   Smart cities definitions – citizens driven

Definition	Reference
'Any adequate model for the Smart City must therefore also focus on the Smartness of its citizens and communities and on their well-being and quality of life, as well as encourage the processes that make cities important to people and which might well sustain very different – sometimes conflicting – activities'	[127]
'A Smart City is a city well performing built on the "smart" combination of endowments and activities of self-decisive, independent and aware citizens'	[4]
'A smart city means *smart citizens* – where citizens have all the information they need to make informed choices about their lifestyle, work and travel options'	[a]
'Smart city initiatives provide opportunities for social inclusion and equal participation'	[128]
'A city that gives inspiration, shares culture, knowledge and life, a city that motivates its inhabitants to create and flourish their own lives'	[129]

[a]http://www.manchesterdda.com/smartcity

refers to instrumenting the city to collect real-time data from both physical and virtual sensors. Once the data is acquired, it can be interconnected across multiple processes, systems, value chains and stakeholders. These two principles effectively promote the physical-digital integration of city systems, allowing cities' systems to be improved and optimized. This is the intelligence principle.

Within this technology, stack intelligence moves out of applications and enters the domain of data: the meaning and relevance of raw data become information that can be provided just-in-time to enable real-time response. Data and modern digital technologies, however, do not lead automatically to smart cities' solutions and services for citizens. Urban systems require open innovation models that enable open, private, commercial and crowd-sourced data as well as people-driven innovation models to transform city data and technologies into innovative services. The Living Labs and other participatory innovation models which emerged from the intelligent cities movement interconnect the current technology push of the technology industry and the application requirements of smart cities [137]. Data and technologies, however, do not lead automatically to new solutions and new services for citizens. The open data urban system demands open innovation models and people-driven innovation models to turn capabilities offered by data and technologies to services and solutions.

Dirks *et al.* [2] stress the importance of the interrelationship between smart cities core systems (e.g. education, transport, waste and recycling, water, energy, health care, buildings, physical infrastructure, food, public safety, etc.). City systems cannot operate in isolation as one is dependent upon the other. For instance, water is essential for energy provision, which in turn is essential for transportation which its demand is increased by greater commerce. The smartness of cities does not only rely on the infusion of intelligence into each city subsystem alone but on the organic whole and how the various systems are connected. In a smart city, the physical infrastructure is infused with information to allow a further increase in the efficiency of existing business, enable the creation of new businesses through integration, improve people's mobility, mitigate environmental changes, recover rapidly from disasters, collect and exchange data across entities and domains to make better decisions. From this perspective, contributions to urban competitiveness are also associated with the possibilities opened by ICT fostering an integral approach to urban development and management.

3.1   The physical-digital integration of smart cities

Cities are composed of many fragmented information systems and applications which provide services for citizens, businesses and public departments (Figure 3.1). Among these fragmented systems are included geographic information systems, intelligent traffic systems, health care, buildings and constructions, security, transportation, water and waste supply, environmental monitoring systems, energy and electricity. In addition, cities are also equipped with heterogeneous devices and have different stakeholders.

These fragmented systems are often developed/operated using different languages and data formats, run over different networks, infrastructure and environments, varying in terms of architectural design and communication protocols, are geographically distributed, and according to Rotmans [138], they also exhibit emergent behaviour.

These are the main characteristics of complex systems of systems (SoS). Kotov [139] defines an SoS as '*large scale concurrent and distributed systems that are composed of complex systems*', and Lukasik [140] complements this definition as '*the integration of systems that ultimately contribute to evolution of the social infrastructure*'.

For Sage and Cuppan [141], an SoS exhibits the following characteristics:

- Operational and managerial independence.
- Geographical distribution.

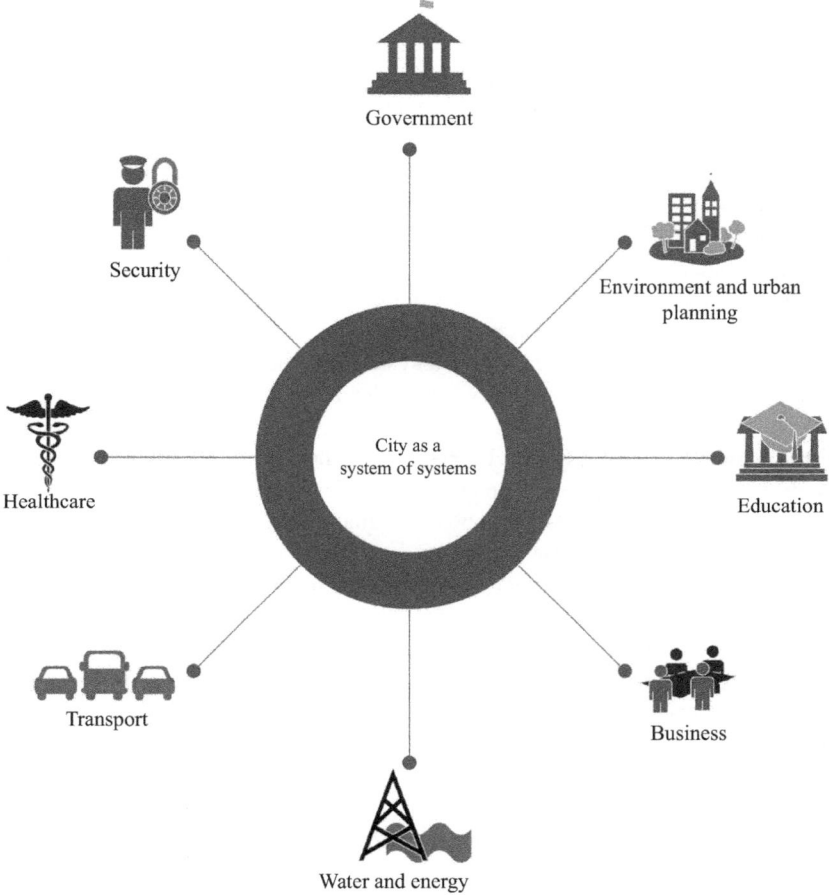

Figure 3.1   City as a system of systems

- Emergent behaviour.
- Evolutionary development.
- The systems shall function holistically through SoS defined interfaces and performance parameters to achieve a new level of performance and capability.

Selberg and Austin [142] complemented these definitions of SoS with the addition of two more principles:

- The complexity of the SoS framework does not grow as constituent systems are added, removed or replaced.
- The constituent SoS does not need to be re-engineered as other constituent systems are added, removed or replaced.

Rotmans (2006) describes cities as complex adaptive systems with seven principal properties, of which we highlight:

1. The cause-and-effect relationships among city components are non-linear; either negative or positive feedback loops can be found in the city system.
2. Energy and information are in constant flow across system boundaries as the system is open.
3. The structure of the system is diverse and encompasses a variety of interacting components.
4. The system has multiple attractors – preferred system states – where the system is moving towards a specific attractor.
5. City systems are themselves complex system creating an overall nested structure, and the relationship between city system leads to the emergence of patterns.

As city systems, processes and services have not been properly interconnected; immediate reaction in response to emergencies and the efficient management of urban resources are limited. For example, one of the main challenges found in the management of the urban environment is how to promptly react to unpredicted emergency situations. Often, immediately real-time reaction can be limited because of the fragmented scenario found in city systems, lack in connecting sensors and actuators and business processes and services which have to change in order to allow immediate response to the emergency. Such an ecosystem is hardly found in a state of equilibrium. Most important is the high dynamics of the changes that can occur in the city environment due to an emergency situation.

Modern digital technologies have the potential of unlocking unprecedented opportunities to integrate city fragmented systems and to optimally and transparently manage and operate the city infrastructures. While for the last decades legacy systems have been primarily designed for specific purposes with no/or limited flexibility, modern digital technologies enable applications and service platforms to capture, communicate, store, access, extract insights from and share data from the physical world. In the context of smart cities, there is an increasing demand for novel applications and services which are able to provide anytime and anywhere access.

As cities seek to reduce costs and interconnect their activities, their systems environments have increasingly been tied to a pool of semi- and non-structured real-time data which is catalysed by millions of electronic networked devices responsible to manage and operate the city infrastructure (e.g. sensors, smart meters, cameras and actuators). Once real-time data is collected, the cities physical infrastructure is infused with information, and the data can be integrated and exchanged across multiple processes, systems and stakeholders. These two steps realize the physical-digital integration of city systems that allows cities' applications, machines and human users to better understand their surrounding environments. The integration of city systems at the system-of-system level has been demonstrated to be able to create drivers for infrastructure innovation and improve the control of resources [22].

The development of smart cities has been mainly driven by emerging technologies, social engagement, big data and open data initiatives. In the following sections,

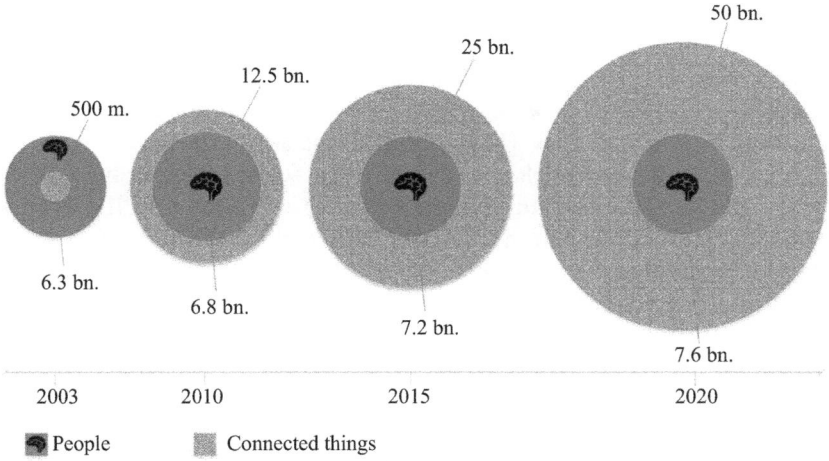

Figure 3.2   The dawn of connected machines

we discuss the physical-digital integration of city systems and its main technology enablers.

3.1.1   The Internet of Things

The Internet of Things (IoT) emerged at MIT around the year 2000 [143] and it is defined by the International Telecommunication Union (ITU) as '*a new dimension has been added to the world of ICTs: from anytime, anyplace connectivity for anyone, we will now have connectivity for anything. Connections will multiply and create an entirely new dynamic network of networks – an Internet of Things*'. The industry sector[1] predicts that by 2050 over 50 billion devices will be connected to the Internet, representing 6.58 devices per human (Figure 3.2).

Cash *et al.* [144] define M2M as the association of computing technologies and communication, via smart and connecting objects that interact without human intervention, with the information systems [145]. These physical and pervasive objects have identities, physical attributes and virtual personalities, use intelligent interfaces and are currently referred as '*things*' [16]. These things have identities, physical attributes, virtual personalities, use intelligent interfaces and can interact and collaborate within their environment using unique addressing schemes in order to achieve common goals. Besides smart objects, machine-to-machine (M2M) communication is another fundamental component of the IoT. M2M is a low power and low-cost association of computing technologies and communication, via smart and pervasive objects that interact without human intervention, with the information systems [145].

---

[1] https://www.cisco.com/c/dam/en_us/about/ac79/docs/innov/IoT_IBSG_0411FINAL.pdf

This environment composed of connected things has created an entirely new dynamic network of networks which has revolutionized the world with the idea of the connectivity of anything, from anyone, at anytime and anywhere. The IoT is '*a world-wide network of interconnected objects uniquely addressable, based on standard communication protocols*' [146]. In the IoT paradigm, things can be sensors, actuators, lifts, lights, mobile phones, wearables and so forth. Among many of these things, technologies capable of tracing and addressing items such as RFID and sensors networks play an important role in the IoT. RFID, in particular, still stands at the forefront of the technologies driving the development of IoT, especially for its maturity and low cost which has made it widely adopted in the business community [147].

In addition to wireless communications, memory and elaboration capabilities, these things are often equipped with some capabilities such as autonomous and proactive behaviour and context-awareness. Research in this area has gained increasing attention from academia, industry and standardization bodies (e.g. telecommunication, semantic web) in the past few years. This demand is a response to the mostly needed adjustment on services (e.g. city, enterprises) in order to accommodate the limitations of mobile and ubiquitous computing (see [148]) and take advantage of user's digital take-up and the features which are being offered by new smartphone technologies. Some of the key drivers of the IoT are [149]:

Moore's law: 'The number of transistors on the same chip area doubles every 18 months' [150]. Sensors are becoming cheaper and smaller and are increasingly being embedded in common daily settings, creating connections among people and things.

Advancements in telecommunication technologies: Ever more the society is relying on the use of computers/mobile communication to perform tasks of the daily life. Advancements in wireless communication, intelligent transport systems and 5G expected to further develop the digital communication process [151]. Such advancements in 5G wireless communications technology [152] and network communications will provide additional enabling connectivity and increase in data rate that will enable the IoT to be formed by an extensive variety of powerful end-user, network edge and access devices such as smartphones, smart home appliances, connected vehicles, wearable devices, cellular base stations and edge routers, among others. Research on The Tactile Internet [153] is advancing technologies that capture and reproduce various stimuli (e.g. sight, hearing, touch, smell and taste) from the outside world and let humans, as well as machines, perceive and react to the combined stimuli in various ways. These new technologies have a huge potential to be part of the IoT and data collected from such technologies provides unprecedented opportunities to solve social challenges in smart cities.

Fog Computing: Services integration and real-time response in urban environments will require the deployment of massive numbers of sensors which provide location awareness and low latency. The data volume originated from these devices is expected to exceed the design specifications and limits of existing networking systems and today's cloud and host computing models. For instance, health-monitoring, emergency-response and time-sensitive control functions for cyber-physical systems, and other latency-sensitive applications, the delay caused by transferring data to the cloud for storage and processing and back to the application may require prohibitively high network bandwidth impairing scalability or may sometimes be prohibited due to regulations and data privacy concerns.

To date, much of the research efforts on IoT have been concentrated mostly on wireless sensors, cloud connectivity, big data analytics and mobile apps. One of the most challenging open issues in the field of IoT is the efficient use of underlying resources which can directly impact data consistency, reliability and scalability of applications. To address the efficient use of underlying resources, edge computing [154,155] has been proposed to use computing resources near IoT sensors for local storage and preliminary data processing. Fog computing is capable of both coping with latency-sensitive applications at the edge of network and latency-tolerant tasks efficiently at powerful computing nodes at the intermediate of network. Above the fog computing layer, traditional cloud computing with data centres can be used for highly intensive data processing and analytics. Although fog computing approach has the potential to decrease network bottlenecks and accelerate analysis, edge devices, however, have not demonstrated to cope with multiple IoT applications competing for their limited resources. This issue may result in increased processing latency caused by resource contention, and consequently, undermining the potential of time-sensitive applications in smart cities.

The concept of IoT has the potential serve to create a data environment within cities, in which things can capture and communicate real-time data from the physical world. Allowing easy access and interaction with a wide variety of devices in the urban environment (e.g. sensors, actuators, appliances, vehicles, displays, appliances, mobile phones), the IoT will foster the development new integrated services to citizens, businesses and public administrations. Societies are even more reliant on the use of computers/mobile communication to perform tasks of the daily life. Advancements in wireless communication, intelligent transport systems and 5G connections [151] are expected to enhance the digital communication process. Sensors are becoming cheaper and smaller and are increasingly being embedded in common daily settings, creating connections among people and things. Once data is collected from the city, physical infrastructure is infused with information, and the data can be integrated and exchanged across multiple processes, systems and cities' stakeholders.

An increasing number of IoT applications have emerged recently, and they can be categorized as described below [146]. The IoT is the result of the convergence of these different visions.

- **Things oriented:** where the focus is on identifying and integrating objects.
- **Internet oriented:** where the emphasis is on establishing an efficient connection between devices and exploring the applications of IP protocol.
- **Semantic oriented:** which is concerned with describing objects and managing the huge amount of data provided by the increasing number of IoT objects.

However, such a heterogeneous field of application makes the identification of solutions capable of satisfying the requirements of all possible application scenarios a formidable challenge. This difficulty has led to the proliferation of different and, sometimes, incompatible proposals for the practical realization of IoT systems. Therefore, from a system perspective, the realization of an IoT network, together with the required back-end network services and devices, still lacks an established best practice because of its novelty and complexity. In addition to the technical difficulties, the adoption of the IoT paradigm is also hindered by the lack of a clear and widely accepted business model that can attract investments to promote the deployment of these technologies in smart cities [156].

Citizens will be responsible for a great part of the huge volume of data expected to arise in cities. This data will mostly come from weblogs, social media, mobile devices, automobiles, smart cards, smart homes (Home Area Network – HAN) and others. A typical HAN can connect devices such as computers, printers, streaming clients and others. Humans are very powerful devices in themselves and – in most cases – have sensing, actuating and computing capabilities that go well beyond pervasive devices currently available or yet to be invented [157]. For sensing, it is a matter of fact that there are situations and events that only human sensitivity and experience can recognize while pervasive devices have limited capability for such tasks (e.g. quality, interpretation of complicated contexts). To enhance sensing and computing capabilities of our urban environments, users should play an active role and should contribute their own sensing and computing devices (e.g. cars, mobile phones) [158].

The deployment of pervasive technologies allowed a massive increase in the volume of urban mobility data (digital footprint) which has stimulated researchers around the world to understand human behaviour and movement in cities. Ratti *et al.* [159], Reades *et al.* [160,161] and Gonzalez *et al.* [162] used cellular network data to study city dynamics and human mobility, while McNamara [163] used data collected from an RFID-enabled subway system to predict co-location patterns amongst mass transit users. Previous research [164] have analysed over 11 million records from the London's Oyster Card electronic ticketing system to extract patterns of how Londoners travel and explore the regularities in these, linking these movements to the polycentric nature of the centres that compose the core of the city. It is becoming increasingly relevant to explore ways in which such big data can result in tools for people taking decisions within the city.

Smart cities demand applications and service platforms which can capture, communicate, orchestrate, store, access and share data from the physical world. Advancements in data analytics and business intelligence systems can provide the means to visualize and mine information in very large databases, identify patterns and events, reason about such patterns and events, generate alerts and predict trends in cities. Using supercomputers and powerful data analytics tools, city data can be mined and turned into knowledge to increase service levels. Furthermore, merging data from different processes, domains and stakeholders can give insights on how systems influence each other, providing insights on how to increase the efficiency of the overall infrastructure.

3.1.2 The city data

Today many areas of science are facing hundred- to thousand-fold increases in data volumes when compared with just one decade ago [165]. This data deluge is originated from satellites, telescopes, sensor networks, accelerators, supercomputers, etc. We are now generating more data than at any point in the past. The 'digital universe' is expected to reach 2.7 ZB in 2012 and 8 ZB by 2015. The voluminous amount of data created in cities today have been driven by the decreasing cost in computer power and storage, and the proliferation of wireless and sensors networks and connected devices which have started to be embedded into everyday settings and linked with commonplace tasks. As a consequence, 90% of the big data in the world today has been created in the last two years alone [166], and more data crosses the Internet every second than what was stored on the entire Internet in the last 20 years.

Big data refers to data collections (datasets) whose volume exceeds the capacity of database software tools to capture, store, manage and analyse [167]. The definition of the big data volume is subjective and it is not defined in terms of being larger than a certain number of terabytes. Back in 1986, Teradata shipped the first parallel database system (hardware and software) with 1 TB of storage capacity. At that time, that was big data [168]. Today, large data warehouses have many parallel databases and store tens of petabytes of data. There is, however, no specification on the size of big data, that is, how big a dataset should be to be considered as big data [167]. This is based on the assumption that the size of datasets which are qualified as big data will further increase as technology advances. Moreover, they argue that the definition of big data varies by sector and depends on the sort of software tools and dataset sizes that are common in a particular industry.

In this definition of big data, the word '*capture*' refers to the velocity of the data, that is, the data is too fast and must be processed quickly. For instance, fraud detection or emergencies in cities must be identified in a very short period of time in order to allow immediate action. '*Store*' refers to data that cannot easily fit into an existing processing tool. '*Manage*' refers to the need of pre-processing the data to extract useful information, and '*Analyse*' refers to the kind of analysis that existing tools cannot instantly provide. Besides these four challenges of big data handling, data sharing also becomes an issue as there are multiple stakeholders, each one with different interest, objectives and expectations in respect to the big data.

Although many organizations and cities have access to plenty of information, they often find it hard to get value out of it as most of the data is raw form or in a semi-structured or unstructured format [167]. The authors also highlight the great potential and value of big data and cite the US health-care systems in which big data has a potential annual value of $300 billion. In IBM's white paper 'Smarter Cities on a Smarter Planet' published in 2010,[2] the authors describe how cities can be empowered by the efficient use of big data:

---

[2]http://www.ibm.com/smarterplanet/uk/en/

In the past, our data models on the future predicted long-term trends over days, weeks, and years. Now, because we can manage so much data, and we can process it so fast, we can look at traffic and predict what it will look like in an hour. We can look at weather patterns that are coming in and predict where we are likely to have fallen power lines within a square kilometre.

The main technology issue brought by big data is the cost involved in using existing data management technologies which commonly trade off consistency and integrity for velocity and flexibility [169]. The authors define big data as *techniques and technologies that make handling data at extreme scale affordable*. Although there are many new and evolving methods and solutions to manage big data, just applying big data solutions to unstructured and semi-structured data creates a new problem of producing silos of data (structured and non/semi-structured), each with their own limitations and benefits. In fact, the problem is not only deciding which database model is more suitable to use, but instead, the problem is about identifying the processes and activities necessary to integrate orchestrate data.

3.2   The two categories of smart cities' applications

Systems theory helps us to understand the distinction between two different approaches being adopted to design systems to manage and coordinate the provision of city data in an urban environment. City data management solutions comprise a spectrum ranging from closed-system solutions and open-system solutions. Closed systems are those that are not influenced by external factors and therefore less likely to have disruption caused by the environment, which is often unpredictable in nature. Such systems can be controlled and planned centrally. In contrast, open systems take inputs from their environments, transform them and adapts to environmental disturbances when necessary, and then return them as some sort of product back to the environment [170].

3.2.1   Closed-system approaches

Closed-system solutions refers to solutions implemented to integrate city systems in which the data manipulated is not publicly accessible and the entire functionality of the system is determined by a single supplier. It includes specific research and solutions that are centrally controlled and excludes the public outside the organization boundaries as part of the data processing system. In this case and in many others, just a small fraction of data gathered is publicly visible and it is still human controlled. Such solutions have the potential for users within the organization to easily share information, analyses data, and hence contributing to increasing organizational value. The successful implementation of this kind of solution requires governments/businesses to carry out an extensive business change process and technological changes to align organizational processes with the provided solution.

The literature on closed-system solutions for urban data management shows that this area of research is very diverse, and most of the proposed architectures remain in the conceptual phase with few practical and realistic validations. Below, we present some of the most notable work on the management of city data. We present solutions that range from the solution for a specific problem in smart cities, solutions based on data analytics, and general architectures.

Several papers have proposed an array of solutions to solve a specific problem in smart cities and to improve the performance and security of such systems. Examples of such solutions are architecture for video sensing in cities [171]; management and orchestration of heterogeneous sensors for public spaces monitoring [172]; centralized operational central of smart buildings [173]; cost-efficient use of electricity within smart infrastructures and data centres [133]; real-time architecture for disaster management based on information collected from various multiple and distributed sources [174]; physical and technical countermeasures against the security threats [74]; monitoring of network condition of smart cities [175].

Other solutions are more focused on sensor networks, the IoT and the Internet of services. Notable work includes Hernandez *et al.* [176], who proposed a ubiquitous sensor network for which services can be developed at minimal cost. It enables the integration of heterogeneous and geographically distributed sensors in a centralized hub. Andreini *et al.* [177] proposed a service-oriented architecture for M2M interactions, in which objects can publish their services to be accessed by mobile phones. This architecture uses techniques to improve the scalability of the system and rapid recovery of services for smart cities' applications. Live Singapore project [178] is supported by modern pervasive technologies to monitor urban environments and assist in the decision-making process. Anthopoulos *et al.* [179] proposed an enterprise architecture which sets up the information, technologies, specific processes and a framework needed to implement an architecture for data management.

The Smart Santander project [180] created a laboratory for prototyping and developing technologies in a real environment, which contains 20,000 experimental sensors that capture information from different services. Living PlanIT urban operating system designed a real-time control cloud-based platform, which incorporates sensing, simulation, analysis services and applications, enabling – at least in theory – the creation, monitoring and improvement of smart cities. Perera *et al.* [181] give a comprehensive overview of the sensing as a service model and its applicability towards smart cities in the IoT paradigm. Kakarontzas *et al.* [182] suggest suitable architectural patterns that provide a conceptual application architecture framework for smart cities. The architecture patterns are suggested to address interoperability, availability and recoverability, authentication, authorization and confidentiality, usability and maintainability. However, these patterns are very general and have not been validated in a real case study. The point raised in this work and also in the work of Anthopoulos and Fitsilis [179] is the need to adopt a layered architecture to realize smart cities. This style of architecture has been widely discussed in the literature and used in numerous enterprise software and smart cities, and most of the proposed architectures in the smart cities literature are still in the conceptual phase with few practical validations.

Corporations' approach to city data management has also been reported in the literature. Cisco solution for smart cities smart + connected communities on developing cities from scratch and how to instrument these cities using Cisco network solutions and services. The landscape of Cisco as a provider of smart city solutions is to establish themselves as suppliers of network technologies and platforms to integrate systems from a network perspective. This solution is based on the Cisco Service Delivery Platform and represents a kind of operating system for delivery of services. This infrastructure is featured with an application abstraction framework to enable quick integration of enterprise applications.

IBM initiative for smart cities, the so-called Smarter Planet, focus mainly on transforming business processes through data analytics [130]. The architectural design of smart city proposed by IBM is based on the service-oriented architecture paradigm for cross-domain integration and data analysis. The focus of IBM solutions for smart cities (e.g. efficient buildings, water and waste management, energy, city services) is on the SOA layer which provides the data analytics to city managers infer intelligent and strategic decisions. The company is currently working with several cities to establish themselves as the provider of the most intelligent city management solution. A real-world implementation of this approach is the Intelligent Operations Centre built by IBM for the city of Rio de Janeiro, a $14 million facility which integrates data from weather stations, traffic cameras, police patrols, sensors and social-media postings. The entire infrastructure is managed and operated by a single operation centre, allowing managers to visualize cities' dynamic performance in real time.

3.2.2   Open-systems approach

Open-system solutions refer to the removal of the organizational boundaries and the public, which was just an outsider before, now is part of the data processing system and might process, enrich, share, combine, input their own data (e.g. mobile phones, personal devices), and even create high-level streams of knowledge through crowdsourcing. As the size and complexity of open data sources continue to grow, open-system solutions are becoming increasingly ubiquitous. Smart cities' solutions based on open systems are generally more flexible than closed systems because the rich urban data environment it creates enables data sources to be interconnected across multiple processes, systems, value chains and stakeholders.

The US, the UK and the EU strongly promoted the data opening policy, various types of Web 2.0-based technologies are used, including raw data download, open API and linked open data, to provide the data for people and machines to consume. Many smart cities' projects based on open data solutions have increasingly being architected using linked data. Linked data is expressed in open and non-proprietary formats and modular, and systems can be designed independently and later linked at the edges allowing interoperability to be added incrementally when needed, and data can be exchanged and reused across different systems and stakeholders [183]. In 2008, the US government began to publish machine-readable datasets in RDF format, expecting it would help people to easily consume data to create added value and improve government transparency.[3] In January 2010, The UK government opened the public information opening site (www.data.gov.uk) to create societal value by sharing and

[3]The White House Office

utilizing information in the public sector. The UK government serves more than 3000 linked data sets, including house price trend service, private hospital location search and recommendation service; and distributed research and investment information. The Greater London Authority has launched the London Data Store which has made available a large number of datasets from many different domains. Transport for London (TfL) has released several datasets and live feeds for users/developers build real-time applications/systems.

The Open Cities project[4] aims to validate how to approach open and user-driven innovation methodologies in the public sector towards smart cities services in major European cities. The main objective of the project which started in 2011 is to leverage existing tools and infrastructures, trials, prototyping and crowdsourcing, open data and open sensor network platforms. The CitySDK[5] initiative proposes the development of a smart city ecosystem through large-scale citizen's driven pilots that align smart city applications to open-source service developers toolkit. Another smart city initiative driven by the concepts of openness and citizen's participation is the Open and Agile Smart Cities (OASC) initiative.[6] This project aims to incentive the adoption of open standards and principles, thereby facilitating interoperability between different systems within a city and across multiple cities. The smart cities platforms delivered by the OASC initiative is powered and developed by the FIWARE Platform's APIs[7] and CitySDK data models.

Research on the application of linked data in smart cities includes the Open Street Maps project which provides user-generated street maps that is free to use and editable [184], the Urban LarKC [185] and the Traffic LarKC [186] which are integrated with DBpedia [187] are applications which make possible answering queries such as which are the modern art exhibitions that I can reach today in less than 25 minutes if I can get into my car this afternoon at 4 p.m.? [188]. UrbanMatch [186] is a game proposed to use human computation to validate the quality of automatically created links between points of interest and their respective photos. It takes into consideration the features of geo-spatially related Linked data and the knowledge of the players.

Other platforms for city data integration and orchestration includes Attwood [189], who proposed a framework to support real-time analysis, visualization, and modification of a failing system within a smart city. The authors consider critical infrastructures as very important entities, which provide services for maintenance of the well-being of the populace and their failure could cause harm or great disruption (e.g. telecommunication, electrical grid). This framework considers linked-data and semantic web as a key enabler to data interoperability among entities. Le Phuoc *et al.* [190] proposed a software middleware to integrate and share urban data streams using Linked Data technologies. The limitations of their approach is that it is centralised, manual and do not provide ways to track data provenance nor it provides mechanisms to ensure privacy. A more complete approach for integrating urban data using linked

---

[4]http://www.opencities.net/content/project
[5]http://www.citysdk.eu/
[6]http://oascities.org/
[7]https://www.fiware.org/

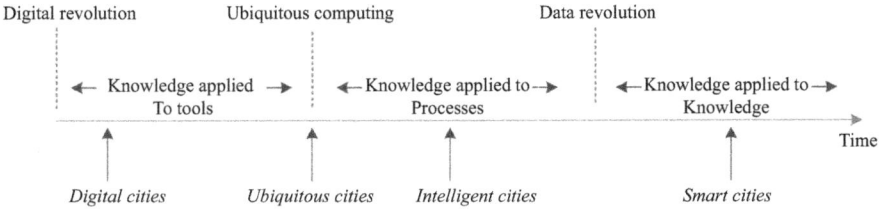

Figure 3.3   *The knowledge framework applied to the periods after the Digital Revolution*

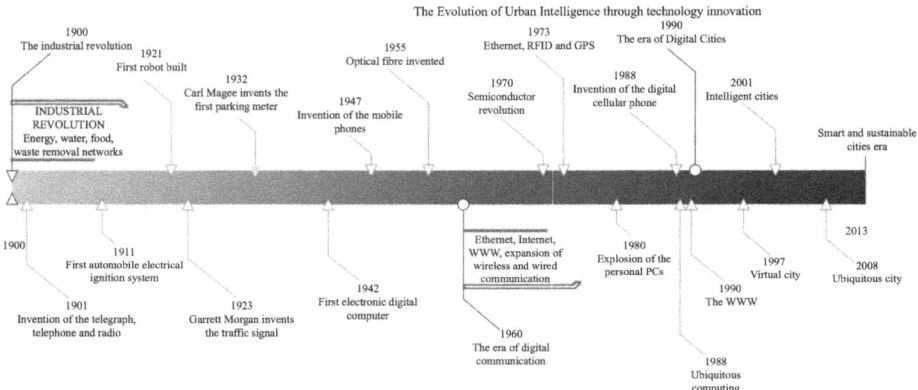

Figure 3.4   *The evolution of urban intelligence through technology innovation*

data has been proposed by Lopez *et al.* [191]. The authors proposed a platform for the city of Dublin in which data can be automatically converted to RDF format and integrated into a local database. This RDF database is linked with DBpedia and with Integrated Public Sector Vocabulary to use a common vocabulary.

Considering the knowledge framework defined in Figure 2.1, we note that at the start of the Digital Revolution, cities started to use technology as a tool to reduce bureaucracy and to facilitate the access of governmental information to citizens, as well as to create a virtual representation of the city. A second wave of revolution appeared with the advancements in digital technologies and the emergence of ubiquitous and pervasive computing. It gave rise to intelligent spaces and cities. At this time, knowledge was applied to tools and to processes in order to understand how to interconnect and integrate fragmented technologies together in such a way that urban processes and spaces can be optimized. The current stage cities are is the stage in which knowledge must be applied to generate actionable knowledge in order to optimize processes and urban services. Figure 3.3 illustrates the knowledge framework applied to after the period of the Digital Revolution, and Figure 3.4 illustrates the evolution of urban intelligence through technology innovation.

3.3   Sources of barriers in city data management

For all smart cities solutions, the overarching goal remains the same: the centralized capture of data produced by all of a city's connected devices and the application of advanced analytic techniques to the enormous volume of data that results in order to realize smart cities. Most of these solutions positioned technology as autonomous and self-contained. What we see in some of the existing solutions is a set of specific technological deployments of components from specific vendors, with their own technological approaches, standards and policies. This implies that each solution/application may interoperable on its very own and original setting and stakeholders, and its component systems may not be linked to larger strategies. As a result, data is not able to be brought together to solve urban challenges.

Our research in the topic demonstrates that the non-technology elements such as business models, value networks, feedback loops, a data marketplace, clear licensing models and more efficient data governance are as crucial as the ones from the technical domain such as data interoperability and open standards.

The following sections summarize the areas with significant limitations and open challenges which have been identified in the current approaches devoted to tackling the problem of data management in smart cities.

Neglecting users needs: Janssen [192] was one of the first researchers on the topic of Open Government Data to demonstrate that one of the main adoption barriers of open data is the users' frustration at the existence of too many data initiatives and lack of ability to discover the appropriate data. Often, the data shared in these 'open data stores' also presents different formats, inconstant quality, are not integrated with any other sources of data on the Web, do not present a well-defined semantics/metadata. As one of the core components of smart cities is M2M data transmission, realising a smart city requires data to be represented in such a way that it is machine readable and the relationship between data is understandable.

The users of city data have different competences in data analysis and manipulation. Therefore, it cannot be assumed that the public will have the same amount of knowledge and competences as researchers and data specialists. Thus, it makes necessary to lower the knowledge required for data access, that is, provide means to people to easily discover and share data, in order to achieve a large-scale dissemination. Research has demonstrated that an e-infrastructure may facilitate faster, better and different scientific research competences and data usage [193]. Such an infrastructure should provide means for data discovery, assessment, provenance, analysis and sharing. In order to enhance the data access, integration and dissemination in an urban environment, cities' stakeholders should be provided with a platform to easily discover all types of city data.

Lack of a data management strategy: Current technology is unable to respond to the increasing demand for integrated city data. The widespread nature of the Internet and the ease with which files and data can be copied, transformed and distributed has made it increasingly difficult to determine the origins of data. The term data provenance refers to the process of tracing and recording processes that led to the creation of a resource and can provide additional evidence for accuracy and timeliness of the data [194]. Goble [195] presents the 7 W's of Data Provenance: Who, What, Where, Why, When, Which and (W)How. Each of these provides a unique type of provenance information which can be used individually or in combination with others to support the assessment of trustworthiness and quality of a piece of data. Goble presents some notable uses for provenance:

- *Reliability and quality*: Given a derived dataset, we are able to cite its lineage and therefore measure its credibility.
- *Justification and inspection*: Provenance can be used to give a historical account of when and how data has been produced. In some situations, it will also show why certain derivations have been made.
- *Re-usability, reproducibility and repeatability*: A provenance record not only shows how data has been produced, it provides all the necessary information to reproduce the results.
- *Ownership, security, credit and copyright*: There are trust and privacy issues not addressed in the data supply chain. Provenance provides a trusted source from which we can procure who the information belongs to and precisely when and how it was created.

In the current so-called 'smart' cities, the existing structures are taken as a starting point, and the user needs for finding, processing and using data are neglected. Among the adoption barriers of city data are:

- *Data findability*: The users' frustration at the existence of too many non-reusable data initiatives and lack of ability to discover the appropriate data on a desired format.
- *Data understandability*: Lack of accuracy and timeliness of data, difficulties in defining data policies and licences and the publication of obsolete and non-valid data which results in the publication of non-value-adding data.
- *Data usability*: Release of data which requires substantial human workload to clean them up for machine processing and to make them comprehensible [192,196–198].
- *Data compatibility*: As researchers now are more inclined to the adoption of linked data, the need for smart cities data integration grows, a problem that is further exacerbated by the lack of widely accepted standards for expressing the syntax and semantics of the city data [199].
- *Metadata compatibility*: Metadata are often insufficient and poorly documented, and therefore users are not able to understand the quality of the data and the way it was gathered and measured [200–202].

Misaligned value network: Another problem commonly overlooked by current solutions is that none of them have considered the full nature of the city data ecosystem and its different data providers: public/private data providers as well as

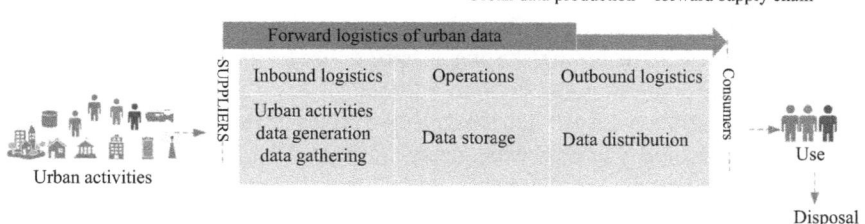

Figure 3.5   Forward logistics of city data

the issues with regard to the usage policies and rights of collected and distributed data. Lack of strong leadership pattern has hindered city data to be exploited to its full effect by all the stakeholders of smart cities. A platform's value to users largely depends on the number of users on the other side of the network, and the value of the platform relatively grows when both sides are matched [203]. Data and services running on city data platforms/portals are often provided and being used by different groups of participants, and the network effects or externalities affect the demand-side economies of scale, meaning that the demand of a service or goods defines its value [204]. Therefore, smart city platforms and data services that do not take into account the role that the stakeholders play in capturing, delivering and generating data is unlikely to achieve widespread approval within a city's communities as the understanding of data privacy and associated problems spread. Furthermore, users should be able to define how their personal data should be used and reused, in a secure way. These actions can drive users acceptance of new data-driven services and products. Open and private data providers in smart cities have evolved from independent units to a dependent network of city data suppliers, and smart cities solutions should consider their strategic roles on the city data platform to survive.

Fragmented data supply chain: The problem with the current smart cities becomes clearer when we analyse the way city data is being produced and distributed. The production of city data involves several stages similarly as physical products produced through supply chains. Considering the logistics network as a large and complex framework of information, in which its success is directly related to the effectiveness in managing information flows, similarly the production of city data can be considered as such, in which stakeholders combine their data and services into complex supply network providing integrated products to their end-users [205]. City data is commonly produced by a forward supply chain (Figure 3.5).

This analogy with supply chains leads to deeper implications when we consider that data, just like a raw resource, can flow through multiple supply chains and end up in a variety of products, making it also vulnerable to breaches. Issues in data trustworthiness, ownership, security, data provenance and legal aspects of data consumption have kept businesses and some data providers out of the open data movement [192,206–209]. Smart city applications – for example, smart meters and smart cards – raise new challenges in data security and privacy since users implicitly expect their data to be secure and privacy-preserved. There is great public fear regarding the inappropriate use of personal data, especially when data is being shared with third-parties.

Another problem arises when the data reaches the last stage of the chain. Essentially, what happens to data when users are done with it? Does it get recycled, reused and upgraded, or the generated knowledge is simply discarded and not reused? *Recycling* of data may equate to maintaining, refreshing and updating data, finding new purposes and applications for it. Likewise, such forward logistics does not offer a mechanism to enhance data quality by collecting feedback from users. The notion of 'feedback' is important in open systems and refers to the situation in which activity within a system is the result of the influence of one element on another [210,211]. The implication of the notion of feedback in systems theory is that, in opening their data, governments should not simply instigate one-way communication of their data but should expect or actively solicit feedback and be able to make sense of this feedback. The opening of systems provides the opportunity for creating feedback loops in which the government can learn from the public. New governance mechanisms, capabilities and processes are necessary for dealing with these feedback loops. The nature of the response depends on the available organizational arrangements that make a response possible [210].

Cities often have different component systems and urban services, and it is reasonable to assume that the same smart city solution will not be replicable in every other city. This is due mainly to the difficulty in generalizing financial, social, technical, environmental and institutional capabilities of every city. McCullough and Dourish [212,213] suggest that technical systems are always given meaning by being situated in a specific local and human context. From the literature review on city data management approaches, a set of common requirements was identified and is presented in Table 3.5. In most of the reviewed research in city data management, the architecture was designed for specific purposes and none of the approaches has fulfilled all of these requirements. This analysis has helped us to understand some of the requirements that must be met when designing and implementing a broader city data management solution.

Although the aforementioned initiatives share such common goals to establish a platform for city data management, there is no consensus on which requirements a platform for city data management must meet to be considered effective for its intended purpose, regardless of how it was implemented. More important is the lack of approaches to guide cities on the identification of their own requirements and needs, and the practical design of their own platforms. Such approaches would give smart cities initiatives a holistic overview of the core domains that go into designing their data management strategy, which would move them towards a more common and aligned set of requirements and innovative business models and processes to acquire and develop data management strategies.

On the one hand, the management of city data involves a range of technical issues, such as bringing technology components of the city together, understanding the

Table 3.5 *Smart cities common requirements*

Requirements	Rationale	Authors
Object and data interoperability	Requirement for the consolidation of any platform that uses a range of objects and data with different formats and semantics.	[172,176, 190,214]
Real-time data	Represents one of the most valuable resources that provides relevant information to be used on the monitoring and prediction of phenomena	[96,172, 181]
Historical data	Consists of a rich pool of information which has been collected over time and may be used to enhance the results obtained from a context and/or data mining and predicting algorithms	[96,172, 181,215]
Data availability	Refers to prompt answers to either real time or historical data requests	[96,172, 181,189]
Privacy	The provision of data services which enable the user to perform their task while preserving their identities and personal information	[181,189, 192,212]
Scalability	The data platforms must be able of '*handling high volumes of data while fulfilling high performance requirements*'[a]	[96,172, 177,192, 215]
Internet of things enabled	Handling heterogeneous sensor data influences the richness of detail and the amount of data that can be processed to produce a wide range of insights	[96,172, 177,215]
Service integration	Services developed must be interoperable so that other services can reuse, group or create a composition using them	[177,215, 216]

[a]EMC2 Documentum Platform, http://uk.emc.com/

semantic challenges introduced by existing data heterogeneity and the nonexistence of a common data model, and eliciting and modelling requirements to design the smart city strategy. On the other hand, it also encompasses a number of non-technical issues, for instance, managing the conflicting expectations of the stakeholders of smart cities with regard to the smart city, understanding societal needs, establishing and designing business models that strike a balance between collaboration and competition among city data and digital services providers.

This increased complexity requires that decision makers are provided with processes and strategies frameworks which assist in the identification of new emerging requirements that will improve the quality of decisions made during the design of data management strategies. In order to improve the quality of decisions, policymakers should be provided with frameworks to handle technical and non-technical concerns and requirements that may arise during the definition made during the design of what we call a **data infrastructure**.

In this book, we highlight the actions needed for the elicitation of the new emerging requirements for data infrastructures in smart cities. These actions are highlighted

Table 3.6   City data management emerging requirements.

Capability	Actions for requirements elicitation
Physical	Identify what ICT infrastructure is needed citywide to support the urban platform
	Map out existing ICT system resources across cities in order to identify those resources with the greatest potential for reuse, identify gaps and provide the foundation for a strategy to fill them
	Develop a picture of the city data landscape, including sources, volume, variety, temporal factors and sensitivity
	Define reasonable service level agreements to guarantee scalability requirements
	Manage the data in a way that ensures its integrity and compliance with data protection regulations
Human	Identify what makes data and services more accessible to users
	Understand under which context users may use or request a service
	Gain understanding of what influences user's experience while interacting with services provided (e.g. usability, feeling of security and trust)
Social	Take human behaviour and needs as seriously as technology
	Understand which services and data are needed to solve social problems and drive innovation
Institutional	Set-up governance processes and usage policies for ICT infrastructure in order to maximize asset reuse by city partners
	Build partnerships to deliver holistic and interoperable solutions that could be rolled out across cities of different sizes
Innovation	Create new partnerships that could allow the creation of new potential and cost-effective beneficial services
	Provide tailor-made data services which careful targeting and needs of users and businesses
Sustainable	Understand which services and data are needed to solve social problems and drive innovation
Economic	Explore use cases where data is used to deliver different forms of value
	Ensure the infrastructure is able to accommodate additional functionality at later stage at a fair and transparent cost

in Table 3.6. We believe that the effective set of requirements will result in a better data infrastructure design and innovative services, which may lead the infrastructure to be recognized as outperforming competitors with regard to city, human and business needs.

Chapter 4

The management of city data

4.1 Current trends in smart cities data management

The systems that operate the infrastructure of cities have evolved in a fragmented fashion across several generations of technology, causing city utilities and services to operate sub-optimally and limiting the creation of new value-added services. These challenges and the scale of city-wide technology adoption have forced cities to rethink their strategies and drive the design of innovative and cost-efficient digital technologies that will provide citizens with a high quality of life while meeting their ambitious sustainability agenda.

With the growing importance of service innovation in smart cities, city data became a more important element in the innovation strategy of cities, which means that more capabilities and resources, including data, have to be made available. In recent years, governments around the world have provided thousands of government open datasets to offer additional data to address complex urban problems. The intensified provision of Government Open Data (GOD) through online platforms has not been entirely voluntary. Many cities, especially in the UK and the US, awoke to it only after being surprised by citizens' and NGOs' demands for increased transparency and accountability.

The UK government has been strongly committed to become the most open and transparent government in the world. The London Datastore was one of the first platforms to make public data open and accessible, and its role was made clear in the Mayor of London's election manifesto 'Boost London Datastore to make the Mayoralty even more transparent and give Londoners access to more information on how I am meeting my pledges'[1] and 'Encourage more partners to use the data on the Datastore to create smartphone apps.' Since its launch, it has published over 600 datasets with open data certificates to assure quality and has led to the creation of more than 200 apps, such as the Citymapper travel app has recently closed a $40 m series B investment and has now been exported to some of the biggest cities in the world, although the ROI and long-term business model is unclear. The London Datastore has been internationally recognized and its success has earned the Greater London Authority the Open Data Publisher Award offered by the Open Data Institute (ODI).

---

[1] https://www.london.gov.uk/moderngov/documents/s11162/The%20Mayors%20Transparency%20Commitments.pdf

Shortly after the first releases of GOD, cities realized that the re-use of their data by private and public bodies became an instrument to foster innovation and to improve and create new urban services. Open data have been used to plan emergency responses [98], to understand how people move and commute within cities [162], businesses can improve their operation and target audiences, citizens can plan their journeys by different modes of transport [163], and innovators and start-ups can access data that will help them to create valuable new businesses. Transport for London (TfL), the leading open data provider in the Transport sector in London, provides over 30 open data feeds to over 5000 application developers who design travel applications, tools and services. Their website receives 600,000 unique views on an average day and 10 million data feed hits a week. Although TfL has no data strategy except to release it, the organization estimates that in 2013 alone their open data initiative generated a value of over 56 million in saved time for users of their services.

Many of the current city data offered in government data catalogue portals are obviously useful. Data produced by services and applications and offered in government data catalogues can be integrated, analysed and visualized. At their best, their flows of data can become the differentiation factor for proving new business models and delivering holistic and interoperable digital services. The integration of city systems at the system-of-systems level has been demonstrated to be able to create drivers for infrastructure innovation on and improve the control of resources [22].

Despite the considerable potential of city data, the process of designing data catalogues, which offers either static or real-time data, has been perceived as complex and cumbersome. City data is provided by a multitude of systems, devices and applications, whose logistical distribution varies according to its suppliers, their sectors, distribution channels and the policies and regulations to which it is subjected to. Each nature of city data must be tackled in a different way, and the data collections that offer opportunities to make cities smarter must be brought together and become part of a coherent and interoperable whole. The diversity, instability and ubiquity of city data make the task of processing, integrating and interpreting the real-world data a challenging task. As it is often difficult to understand the context associated with city data, it is very hard for stakeholders and machines to access and interpret city data unambiguously. Consequently, current smart cities data integration often becomes more about enabling isolated and costly data-driven solutions rather than addressing data integration at the data catalogue level and making the most of standards available for that purpose (e.g. Hypercat[2]).

This confusion may be partially explained by the current global competitiveness on the provision of smart cities; solutions demonstrate the extent to which technology providers and disjoint city-led approaches are seeking to shape up the offering of city data. They are in fact, however, exacerbating the problem of data integration for using their set of specific deployments, proprietary technological approaches, data standards and policies. While some cities have awakened to these risks, they are much less clear on what to do about them and start orchestrating and providing city data through shared and open infrastructures. In fact, the most common city

---

[2]http://www.hypercat.io/

responses to data platforms design have been neither strategic nor outcome-oriented approach but cosmetic: media campaigns and government publications, which are often composed by numerous non-replicable actions or sub-scale pilots/demonstrators and the showcase of fragmented and 'deprived' data re-use.

Existing literature on smart cities and data management rarely offer a coherent framework for data platforms development activities, let alone a strategic one. Instead, they aggregate tales about uncoordinated and disjoint initiatives to demonstrate a city's open data vision. The final outcome is that existing initiatives are susceptible to high development and maintenance costs, information friction, requirements neglecting, non-compliance to data policies and regulations, privacy and data licences violation. These consequences clearly demonstrate the extent to which data platforms are failing the delivery of smart city outcomes and highlight the potentially large financial and regulatory risks for any city whose data management 'conduct' is deemed unacceptable (e.g. misuse of personal data, data license infringement, cybersecurity/cyberterrorism).

The current proliferation of data catalogues has been paralleled by growth in city data ratings and rankings. While rigorous and reliable ratings might constructively influence cities approach towards the provision of city data, the existing self-appointed scorekeepers do little more than add to the confusion. For instance, the Global Open Data Index measures and benchmarks the openness of data around the world on an annual basis. However, the ranking criteria used in the rankings are not quite convincing. The ranking is based on the analysis of openness of 10 predefined datasets only and not on the overall collection of datasets a city catalogue hosts. It weights factors such as the existence of the data, its online availability and formatting. Beyond the choice of criteria and their weightings lies the even more perplexing question of how to judge whether the criteria have been met. 'Is the data machine readable?' is a criterion that is not very easy to measure. Data can be available in digital form but not all digital can be processed or parsed easily by machines.

Finally, even if the measures chosen reflect data openness impact, the data are frequently unreliable. The ODI awarded the London Datastore with the 'Open Data Publisher Award', which celebrates high publishing standards and use of challenging data. However, the great majority of data on the London Datastore has no standardized metadata attributes, is outdated, have neither clear licence agreements nor quality assurance. Often, data is machine readable but not understandable. The ODI and University of Southampton's Open Data Monitor project revealed that of the 218 data platforms (European, national, regional, local) they investigated, nearly 50% of the datasets have no standardized metadata attributes and present 25 different data licence descriptions within their data catalogues. The standards used in data platforms are almost always incompatible with each other, hindering the platform-to-platform integration and the re-use of applications development. In the current moment, such indexes tend to use measures for which data are readily and inexpensively available (not pre-processed), even though they may not be good resources for the smart cities outcomes they are intended to support.

As the complexity of products and service grows, it is increasingly difficult for one organization, no matter how large, to specialize simultaneously in all domains that go into producing and delivering finished products and services [61,207]. Complex products and dispersed services have been increasingly delivered across many firms and markets, making extremely difficult for a firm to innovate alone [207,217]. This creates a greater pressure for organizations to more deeply specialize in their core competence and leave the rest to capable partners. This has increasingly led to the disaggregation of firms into complex supply chain networks involving many partners, and now into even larger ecosystems of smaller firms that specialize deeply in their products and services. Manufacturing, IT services, financial services, engineering and even medical services industries are beginning to see this trend.

Similarly, cities find it tremendously difficult to specialize in all the competencies involved in designing, building and maintaining data infrastructures. However, cities have neither the incentive nor the means of bringing external partners round to the necessary supply chain networks. The consequence of not doing so is that complexity has often overtaken data infrastructures development. Often, it has become longer than a simple ICT project, and as a consequence, more expensive and difficult to design and maintain.

A platform-centric approach for the provision of city data in smart cities enables pooling of multiple organizations' knowledge bases – especially cross-sectorial domains – that are more valuable in combination than in isolation. Building an ecosystem of stakeholders who complements the capabilities of a data infrastructure can potentially bring insights about specialized domains, different application markets, lower costs and shorten the time to market for the development of new services. In the next section, we give an overview of the theory of platforms and their main characteristics.

In an effort to move beyond this confusion, a growing literature on open data and data platforms has emerged, though what practical guidance it offers to governments is often unclear. Examining the prevailing strategy of data catalogues or platforms design is an essential starting point in understanding why a new approach is needed to integrating both technology and non-technology components more effectively into data management and business models.

4.2   A short introduction to platforms

Platforms have emerged as an important concept within different streams of literature – from economics, product development, business strategies and management, as well as software engineering, and the phenomenon has attracted increasing academic interest over recent years [218,219]. Each stream of literature focuses on different empirical contexts of platforms and has different definitions and terms to describe and refer to the concept.

Wheelwright and Clark [220] define the platforms as reduced time-to-market products that meet the needs of a core group of customers but can be extended, replaced or have features removed. Meyer and Lehnerd [221] and McGrath [222] define platform as a system consisting of underlying components that are used in common for the development of dissimilar products within a firm. This is the concept of *product platforms* which is also studied by Krishnan and Ulrich [223] who define a product platform as *'component or subsystem assets shared across a family of products'*.

According to Gawer [218], product platforms are internal platforms and have fixed-costs saving, flexibility in product design and products variety that meet customer requirements and maintain the economies of scale and scope. Economies of scope exist when the cost of joint production is less than the cost of producing each output separately [224], and economies of scale cause the cost of production to reduce as the volume of its output increases [225]. Pine [226] argues that internal platforms are the solution to the facilitation of mass customization, which ensures maintaining benefits from economies of scale. Other key features of internal platforms are the profit maximization for product families [227] and the achievement of familiarity in user experience between different products [228].

A platform provides a foundation upon which complements are built by third parties, which are referred to as the industry [218,219] or ecosystem [218,229,230]. According to Iansiti [229], a platform is

'a set of solutions to a problem that is made available to the members of the ecosystem through a set of access points or interfaces'.

Internal platforms are used in various manufacturing industries, such as the automotive industry [227], airplanes [218] or electronic devices [222]. In the software industry, the use of internal platforms has been adopted in the creation of software product lines. Bosch [228] describes a software product line as an intra-organizational software re-use approach. A software product line consists of an internal platform and separate products that re-use part of the functionality of the platform and extend it with additional functionality.

The wide adoption of internal platforms in industry demanded new strategies capable of enabling the adoption of platforms at an inter-organizational level. The lack in standardization caused high variety of parts or intermediate products that were produced in industry supply chains [218]. In order to address this issue, supply chain platforms were introduced and adopted in industries such as the automotive industry [231] and high-tech industry [232]. One of the main distinct characteristics of internal and supply chain platforms is that they thrive on the contribution of a single or a small group of stakeholders. On the other hand, industry platforms and two-sided markets influence groups of actors. In the economics literature, the terms two-sided market or multi-sided platform are used to describe a product, system, service or even an organization that mediates the interaction between two or more groups of actors [61,233]. Common examples of industry platforms are Microsoft Windows, the iOS operating system and the Apple iPhone, Facebook and Twitter [218,219,234,235]. These examples demonstrate that industry platforms are predominantly found in the software industry. Gawer [218] defines an industry platform as

'A product, service or technology that is developed by one or several firms, that serves as a foundation upon which other firms can build complementary products, services or technologies'.

Typically, in a multi-sided platform, complementary products and services running on top of the platform are offered by different independent providers and do not take ownership of the goods and services whose transactions they facilitate. Instead, multi-sided platforms providers alleviate bottlenecks for buyers and sellers by mediating their transactions with one another and generate value for buyers and sellers through enhanced market efficiency, such as transaction volume, resource allocation and an improved matching between supply and demand [236].

The platform at the centre of the two-sided market is created and maintained by the platform provider and sponsor, who may or may not be the same entity [237]. Platform sponsors, as owners of the platform, hold property rights over a platform, and as such have the right to modify a platform's technology, to determine the design of the platform's components and rules and to determine who may or may not participate in the two-sided market [237]. For the purposes of simplicity, in this book we assume that the platform provider and sponsor are the same entity, which will be termed the *platform owner*.

A platform runs on the top of a shared infrastructure technology and is composed of many applications that interoperate with the platform. Besides these core elements of a platform, there are three other additional elements: the end-users, rival platform and the competitive environment in which they exist. End-users, or stakeholders, are the collection of existing and prospective adopters/users of the platform. A platform exists within a larger competitive environment, in our case the smart cities environment, and they often compete with other rival platforms. Platform ecosystems are used as a metaphor to describe and study the inter-relatedness of different actors that are part of markets, entire industries or complementor networks [218,230,238]. The concept of ecosystems has been borrowed from natural systems [239], and Moore [238] was the first one to introduce business ecosystems as he argued that similarly to natural ecosystems, business ecosystems were self-reinforcing systems that consist of different interdependent species. Translated to the traditional business domain, Moore [238] argues that firms cannot innovate in a vacuum, rather they depend on other parties to collaborate with them. Table 4.1 summarizes the concepts of platforms and their respective definitions.

In the field of software engineering and information systems, researchers use different terms to refer to multi-sided platforms. For instance, Tiwana *et al.* [240] use the term *software platform* and define it as

'the extensible codebase of a software-based system that provides core functionality shared by the modules that interoperate with it and the interfaces through which they operate'.

Hanseth and Lyytinen [241] use the term *information infrastructures* and define it as

'a shared, open, heterogeneous and evolving socio-technical system of information technology capabilities recursively composed of other

Table 4.1   Platforms definitions

Concept	Definition	Reference
Internal platform	'Reduced time-to-market products that meets the needs of a core group of customers, but can be extended, replaced or have features removed'	[220]
	'Internal platforms and have fixed-costs saving, flexibility in product design and products variety that meet customer requirements and maintain the economies of scale and scope'	[218]
Supply chain platform	'A set of subsystems and interfaces that forms a common structure from which a stream of derivative products can be efficiently developed and produced by partners along a supply chain'	[218]
Multi-sided platform	'A product, service or technology that is developed by one or several firms that serve as a foundation upon which other firms can build complementary products, services or technologies'	[218]
Software platform	'The extensible codebase of a software-based system that provides core functionality shared by the modules that interoperate with it and the interfaces through which they operate'	[240]

infrastructures, platforms, applications and IT capabilities and controlled by emergent, distributed and episodic forms of control'.

What is common in all definitions of a platform across economic, business management and information systems research is that platforms are modular architectures in which core independent modules are being used and re-used across multiple products and services [242,243]. Moreover, as products and services running on platforms are often provided and being used by different groups of participants, platforms mediate multi-sided networks, which exhibit '*network effects*'.

Regardless of the term being adopted, the focus of this stream of literature focuses in theorizing the development, governance and evolution of large complex information systems as platforms [244–246], on which new services can be added to benefit from shared data resources [245].

4.2.1   Network effects

Network effects or '*network externalities*' are the interconnection of different groups of users around a multi-sided platform and occur when the value of a platform depends on the number of users or on the number of services or complementary modules on the platform. Network effects arise when the desirability or functionality of a product depends on the number of complementary goods available for it [247]. According to Shapiro and Varian [117], network effects or externalities affect the demand-side economies of scale, meaning that the demand of a service or goods defines its value. Network effects can be direct or indirect.

Direct network effects: Intra-side externalities occur when users value a platform based on the number of users in the same group (e.g. network effects between the users of social networking platforms such as Facebook and Instagram).

Indirect network effects: Inter-side externalities occur when the value of a platform for a group of users depends on the participation of another group of users (e.g. indirect network effects between users of a game console and game developers) [61,248]. Shapiro and Varian argue that the demand for a service or goods defines its value, so the network effects of a platform can affect the demand-side economies of scale. Research has demonstrated [203] that both network effects must be comprehended, as well as different types of costs related to the platform. On the same side, a positive network effect can be obtained by economies of scale, while a cross-side phenomenon brings indirect network effects.

4.2.2   Platform actors and relationships

Actors – or stakeholders – play a crucial role in ecosystems, as they engage in relationships and provide for interaction, exchange and innovation with and among each other. Actors can participate within one or more ecosystems, for example, by developing complements for a platform [219], or shape [229] and orchestrate [230] platforms. Efforts have been made to classify actor groups that share similar characteristics. For instance, Hagel's [249] research distinguish between shapers (e.g. the actors who are involved in the development of the platform) and participants that benefit from the effort the shaper has put into creating the ecosystem, in which the participant can create business value. Iansiti and Levien [229], Iyer [250] and Jansen and Finkelstein [230] employed a comparable classification that incorporates the following roles:

Keystone: An actor who provides the foundation for the ecosystem with its standards, technology or platform. Keystones create value by providing a core technology (e.g. a platform) to be used by other members of the ecosystem, they provide incentives to encourage more participants to join the ecosystem and innovate around the platform and they share the value from the business ecosystem with other members, or providing tools and facilities that benefit members [229]. In the smart city scenario, the keystone is the government/policymakers willing to offer city data to interested stakeholders.

Dominator: An actor who progressively eliminates other actors within the ecosystem through merging and acquisitions. Iansiti's research [229] suggests that dominators' strategies damage the health of business ecosystems by reducing its diversity and competition and hinder the creation of innovation. The platform leadership strategy adopted by smart cities policymakers may hinder the emergence of dominators.

Niche player: An actor who uses the technology, standard or platform provided by the keystone to create business value. Niche players represent the least influential members of business ecosystems; however, as they are specialized in different domains, their presence is essential to ensure diversity around business ecosystems [229]. In the smart cities scenario, niche players are, for example, the ones who publish, integrate and enrich data and provide tools to facilitate data consumption and dissemination.

4.2.3 Platform leadership and governance

Platform leaders are presented with a large set of challenges as they seek to develop and maintain their platform to the upfront of their industry. Nevertheless, research has shown that not every product or service can be turned into a platform [251]. The first reason is that a product can only be turned into a platform if it performs one or more elementary functions within a system. Secondly, the potential platform needs to solve a business problem for multiple actors/stakeholders or organizations. A strong value network of partners has demonstrated to reduce complexity and costs associated with platform development [251]. When organizations establish a platform they strive for platform leadership. Platform leadership is defined as '*The ability of a firm to drive innovation around a particular platform at the broad industry level*' [219].

Platform owners have mostly to deal with technological innovation, platform evolution and the preservation of market leadership [219]. Platform owners must understand how to facilitate continuous compatibility with complements, or how to evolve a platform to prevent it from becoming obsolete. For instance, Gawer [218] describes an initially failed initiative of SAP to position themselves as a platform leader. SAP which was traditionally a business application vendor decided to introduce a platform called NetWeaver that should provide for increasing compatibility with non-SAP products, for which complements should be built by their partners. As SAP did not abandon selling its existing products, SAP ended up competing with its own ecosystem of complementors, who they initially did want to support in their activities. Different forces are at play in establishing and maintaining an industry-wide strategy. There are four dimensions – or '*levers*' – that must be addressed by a successful platform strategy [219]:

Scope of the firm: A platform owner has to make decisions about what is done in-house and what is left to complementors. Platform owners must carefully assess the opportunities that arise to enter complementary markets [251]. For instance, active participation and acquisition of complementors may force other complementors to depart the ecosystem, as they feel the platform owner diminishes the opportunities for complementors [229,252]. Common data services platform owners have mechanisms at their disposal to stimulate innovation by individual complementors in the ecosystem, such as disclosing architectural details, sharing expertise and create partnerships [219,253].

Product technology: Platform owners have to make decisions about the architecture of their platform and the surrounding interfaces and must decide on what information is shared about the platform and the functioning of its interfaces. The choice for certain foundation technology architecture has an impact on the fate of the platform as a whole. The authors argue that inaccessible architectures may hinder the adoption of the platform by complementors, as opposed to a modular architecture with interfaces that reduce entry barriers due to increased transparency and integrativeness [219,247]. Sharing technical information about interfaces will help complementors in targeting opportunities around the platform [248].

Relationships with external complementors: Platform owners have to make decisions about establishing competitive or collaborative relationships with complementors, manage network of complementors, support collaboration and joint innovation, while complementors, at the same time, engage into competitions [230]. Furthermore, it is important to agree about the standards that are used for interaction between complements and platform in the complementor ecosystem. This practice is known as balancing consensus and control [219]. In a data infrastructure, policymakers must ensure that the stakeholders of city data are able to co-exist and cooperate in the platform.

Internal organization: Platform owners have to make decisions about how to design and use their organizational structure to avoid internal and external conflicts of interest. As in co-opetition models, a platform owner may end up with a modular organizational structure. Based on the four pillars of platform leadership, Gawer and Cusumano [219] identify two generic platform strategies, one for existing platform leaders and one for new entrants who want to build a platform from scratch, which they refer to as tipping and coring. *Coring* involves providing complementors with incentives to become part of the ecosystem and showcase their complements. In addition, coring can also provide opportunities for complementors, likewise the concept of niche creation described by Iansiti [229]. Gawer [218] argues platform owners to have a strong protection of intellectual property with regard to platform architecture and to increase switching costs for complementors by coupling them tightly with the platform. *Tipping* involves platform-based competition and focuses on the development of appealing features, driving innovation by complementors, subsidizing complementor development or penetration of new markets through complementary products.

With regard to the first lever of platform leadership, platform leaders should take over responsibilities for an essential part of the platform and ensure that the essential part is easy to connect to and add to so that complementary services can be assigned to external contributors [248]. Choi [89] suggests platform providers to focus and provide parts which are core to their business to ensure a significant contribution and maintaining the business for a long time.

The second lever of platform leadership is about the adoption of platform openness strategies. Platform providers adopt different platform openness strategies to leverage participation of complementary providers while keeping the competitive advantage over them. For instance, one could provide appropriate information regarding rules, standards and APIs of the platform to encourage and assist complementary providers to innovate on the platform. Platform openness is a very critical lever of platform leadership as while strict control may hinder innovation, complete openness intensifies competition and may discourage investment by complementary providers and/or put the platform leader at risk of losing control over the platform.

4.3   The two prevailing approaches in the provision of city data

Broadly speaking, the providers of city data in smart cities have used two main strategies to make their case: to adopt either a bottom-up or a top-down approach. On the one hand, bottom-up strategies to city data provision are independent approaches which are often neither addressing the needs of the stakeholders of city data nor integrating their capabilities to city's larger strategies (the top-down approaches). On the other hand, top-down approaches are city-led approaches which are often neither taking social influence into account nor maximizing the efforts of other initiatives who are working towards the provision of city data (the bottom-up approaches).

Bottom-up approaches: Bottom-up approaches have been widely adopted in confined and disjoint initiatives. Such initiatives address the problem of city data management as a single IT development project which is not fully integrated nor linked together in a way that allows cities to efficiently leverage their collective knowledge.

Often bottom-up approaches struggle to maintain interoperable data that is seamlessly integrated across different systems and stakeholders. London's functional bodies, for instance, encompass the 32 Boroughs, the City of London, the Greater London Authority (GLA), Transport for London (TfL), the Mayor's Office for Policing and Crime (MOPAC), London Fire and Emergency Planning Authority (LFEPA) and the London Legacy Development Corporation (LLDC). All of these groups generate their own data but share little of it with each other. The Infrastructure Mapping Application developed by the Infrastructure & Growth group of the Greater London Authority is one of the emerging public sector solutions challenging this fragmentation.

The private sector is an active producer of data in cities. Private sector data are produced, collected or funded by the private organizations (e.g. telecom, utility companies), which can be either released as open or proprietary (commercial) data. In the case of proprietary data, it is often represented using industry standards, is subjected to charge, usage authorization, licencing agreements, privacy restriction and distribution boundaries which are decided by an individual organization.

As the need for data integration grows, the problem of data interoperability is further exacerbated by the lack of widely accepted standards for expressing the syntax and semantics of the city data. In such environments incapable of reciprocal operation with other, there are high levels of heterogeneity in terms of data, hardware and software elements. Together they become a set of specific technological non-interoperable deployments of components from specific vendors. The implication of this scenario is that each solution/application may interoperable on its very own and original setting. Upgrading this complex environment is time-consuming and has high-cost implications, particularly if there is a high-level use of proprietary and legacy systems (e.g. [254]).

Top-down approaches: Top-down approaches suggest that cities act solely as 'implementers of initiatives' – are prominent in the goal of the majority of GOD strategies. Led by the smart cities '*buzzword*', many cities have assumed that answering to governmental pressures for data release before understanding technology and users' needs for data finding, processing and sharing is the best tactic to follow. That may be a natural reaction, especially coming from cities whose data catalogues have been outsourced to private organizations which are willing to sell their proprietary solutions; however, it can be a dangerous one.

By failing to follow appropriate leadership strategies, many initiatives created to offer city data and support smart cities vision have not been as productive as they could. Most top-down approaches neglect the expectations and needs of the various users of city data (e.g. citizens, businesses, entrepreneurs, data scientists, research institutes, city councils). Data.gov.uk and data.gov are well known for being reasonable proxies for public sector information available to the general public, although certainly not comprehensive sources. As same as the aforementioned bottom-up approaches, the users of these initiatives are not provided with data that can be both human and machine readable and understandable. For instance, nearly 60% of the top ten data formats provided in data.gov.uk are proprietary formats (e.g. .pdf and .xls). A very large number of datasets in both initiatives require substantial human workload to data cleansing and semantics sorting [192,196,198]. Often users are provided with obsolete and non-valid data accompanied by insufficient and poorly documented metadata attributes [200–202]. User's different competences in data analysis and manipulation is an issue commonly overlooked in both top-down and bottom-up approaches. It cannot be assumed that the public will have the same amount of knowledge and competences for city data manipulation as researchers and data specialists. Thus, in order to achieve a large-scale dissemination, it is necessary to lower the knowledge required for data access and provide means to people to easily discover, re-use and share data.

Furthermore, as discussed in previous chapters, the corporatization of smart cities and its consequent vendor lock-in can become a big and expensive concern for city government. Schaffers [255] noted that '*smart city solutions are currently more vendor push than city government pull based*', and given the current favourable market conditions, companies divert funding streams and create public–private partnerships. As such, companies push cities to retrofit their infrastructure with proprietary technologies that ensure they become an indispensible part of and administer [3] how various aspects of city life are created, monitored and regulated.

Although such systems provide control over some critical resources, they become very expensive and hard to customize, maintain and extend. This is due mainly to the closed and proprietary nature of such systems, which are often developed/operated using different languages and data formats, varying in terms of architectural design and communication protocols. This situation has caused cities to spend considerable effort to manage multiple independent systems. Consequently, problems have been reported with achieving full utilization of such systems, and policy-makers and infrastructure providers yet do not have a fully integrated management solution. The lack of interconnecting systems and technologies within buildings limits optimal control of resources and a prompt response to unpredictable events. The most challenging issue is that corporate dependency is hard to be undone or diverted [256].

While governments are increasingly aware of the impact of effective city data and services offering in smart cities, these impacts can be more restrained and variable than many policymakers realize. Data infrastructures depend on the context of the smart city, that is, they depend, for instance, on location, culture, available data, smart city vision, local regulations and policies. The same data infrastructure will have very different capabilities or impacts in different locations. A one-size-fits-all approach to data platforms transformation into data infrastructures and simplistic approaches to engage stakeholders of city data with one another are unlikely to work.

4.3.1   Towards a data infrastructure

The two strategies of city data offering share the same weakness: They focus on the tension between their ambitious targets for data provision and technology rather than on the strategic role of non-technology components in bringing them up together. These approaches even in combination are not enough to tie the strategies of community-led developments (e.g. local authorities, utilities, telecom and private organizations) to major strategies (e.g. city- and national-led approaches). Consequently, none of the existing city data strategies on its own is sufficient to help a city to identify, prioritize and address non-technology and technology issues that matter most of the ones on which it can make the biggest impact in integrating cross-domain value-added city data. The result is oftentimes restricted and uncoordinated city data activities disconnected from the wider smart cities context that neither make any meaningful impact nor strengthen the city's long-term competitiveness in the smart cities arena.

Internally, city data initiatives and practices are often isolated from similar efforts – even separated from their overall vision. Externally, they become diffused among numerous unrelated efforts, each responding to a different stakeholder group or a business model. Taken together, these challenges mean that both isolated top-down and bottom-up data management approaches cannot work. The consequence of adopting such approaches is a tremendous lost opportunity. The power of cities to deliver smart cities outcomes is dissipated and so is the potential of cities to take actions that would support both their smart cities vision and their strategic and economic goals.

However, overcoming these challenges is not a simple task. Today's simple reality is that it is tremendously difficult for cities to specialize in all the competencies involved in capturing, storing, orchestrating, maintaining and distributing cross-domain city data. The tendency of existing studies has been to provide frameworks to develop the ITC infrastructure of cities while disregarding the evolving interactions between users and service providers over time.

Hence, cities can be better served by designing platform-centric data infrastructures which employs the concept of data-driven value networks, leading to the delivery of new services, themselves underpinned by new value chains/networks. It involves adopting platform-based competition and innovation, and negotiating con-flicting requirements of the entire network of collaborators who contribute to the data infrastructure by either providing data or services. Through a more deliberate orches-tration of open and proprietary data provided by both public and private sectors, and address privacy and trust issues in relation to volunteered citizen data, we believe that the current isolated existing initiatives could be transformed into suitable data infrastructures ready for the data deluge that the Internet of Things and GOD will bring about.

Data infrastructures acknowledge the relationships that make the city data supply chain more effective in a way that produces lower costs, better data and services, and which lower risks for each of its participating members.

Part II

The link between data infrastructures and business strategies

Chapter 5
Services innovation and business models

5.1 Context

To advance data infrastructures, we must root it in a broad understanding of the interrelationship between city data offering and business models while at the same time anchoring it in the strategies and activities of specific cities. Although saying broadly that a data platform needs a strategy seems like a very straightforward concept, it is the basic truth and the comprehensive activity that will pull cities out of the confusion that their current data offering approach has created. Successful smart cities need effective data infrastructures. The simple reality is that cities must respond to the increased market demand for a more integrated provision of city data. Ultimately, a true smart city provides value-added data which expands demand for business and innovation in the long term, as more social needs are met and aspirations grow. Any smart city that pursues its ends based on poor quality data in which its infrastructure will operate on will find its success to be illusory and ultimately temporary.

At the same time, data infrastructures need effective business models. Complex products and dispersed services have been increasingly delivered across many firms and markets, making it extremely difficult for a firm to innovate alone [207,217]. This has increasingly led to the disaggregation of firms into complex supply chain networks involving many partners who specialize deeply in their products and services. Regarding smart cities, the provision of city data and specialized services for data manipulation are frequently centrally controlled and excludes units outside the organization boundaries. Nowadays cities, we argue, have neither the tools nor the means of bringing external partners round to the necessary supply chain networks. This is especially due to a number of barriers to joint collaboration beyond organizational barriers, including funding, expertise and the willingness to work across silos. The consequence of not doing so is that complexity has often overtaken the management of city data. Often, solutions to manage city data have become longer than a simple ICT project, and as a consequence, more expensive and difficult to design and maintain. If smart cities are not designed to join up across city data silos to address urban challenges, they will limit the ability city data delivering the smart city promise.

Leaders of city data offering in smart cities have focused too much on the friction between technology components and not enough on the points of their intersection with non-technology components and business models. The mutual dependence of city data offering business models implies that any strategic and technology decisions must follow the principle of shared value, or what we call a middle-out leadership pattern. That is, choices must benefit both top-down and bottom-up approaches. If either top-down or bottom-up approach pursues standards, policies and regulations that benefit only their own interests, it will find itself on a dangerous path. A temporary gain to one will undermine the long-term prosperity of both sides.

Rather than acting solely as *implementers of initiatives*, government initiatives must take social influence into account while maximizing the efforts of other stakeholders who are working towards the achievement of the same goal: to facilitate deep knowledge discovery and the creation of new valuable integrated services through the exploitation of rich, interoperable and engaging cross-domain city. Data infrastructures can provide many functions that transcend space (and time), break down the barriers to information access and enhance communication and collaboration. Thus, it enables people to have access to information that will enable them to innovate, to work better, to commute more efficiently in between places, enable governments to get insights on the urban services being provided anywhere and anytime they want.

To put these principles into practice, our business models framework combines city data offering with a business model thinking to renew and extend common innovation and competitive strategies (e.g. [257–259]), and address intra- and inter-firm issues such as organizational change, value network design and innovation management [260–263]. From a practical perspective, the main purpose of our framework is to allow governments to create, deliver and capture value through data infrastructures which are designed on the basis of social influence and not authority.

5.2   An introduction to business models

Innovators who have the potential to create positive social and sustainable effects need to diffuse beyond the niches in which they emerge to become effective sustainability innovations [264–266]. Smart cities entrepreneurs are faced with this challenge when they try to disseminate new solutions to urban problems through commercial activities and aim for large market shares and socio-political influence (e.g. application development). These smart cities innovators tie their business and economic success directly to the achievement of positive effects for humankind, mobility, urban services and the natural environment. However, current research reveals significant uncertainties related to innovation-centric approaches: "Innovation has been widely regarded as a panacea for sustainable development, but there remains considerable uncertainty about how it will lead to a more sustainable society" [267].

The increasing global competition among countries and citizens demands have requested serious cities to optimize the provision of their services. Modern digital technologies offer a new wave of opportunities to mitigate some of these impacts and create a balance between social, environmental and economic opportunities, and exploit virtual value chains [268]. The current advancements in technology, the decreasing costs on computation power and the miniaturization and increased mobility of devices have changed the way systems operate cities' infrastructures. Systems that operate within urban environments are increasingly being tied to a pool of multi-structured real-time data catalysed by millions of electronic networked devices – the city IoT (e.g. sensors, smart meters, cameras and actuators). Such advancements have increased interoperability between services, security and natural interfaces [269] and have enabled the rise of architectures and platforms for knowledge sharing, collaboration and electronic commerce transactions, anywhere, anytime.

As a consequence of these developments, cities have changed the way in which they interact with their citizens, businesses and service providers by the exchanging of a vast amount of data every day. The users of urban services have increasingly influenced the way such new digital services are created and incorporated into their day-to-day routines. Service innovation is mostly user driven and is about co-design of services through the provision of feedback which will be used to improve the operation of existing urban services [270]. Furthermore, increasing interaction with the users of services has the potential to improve the effectiveness of service innovation as users are an important source of information for service innovation [271].

With the growing importance of services, service innovation becomes a more important element in the innovation strategy of cities, which means that more capabilities and resources have to be made available. Hertog [272] argues that service innovation is based on service concept, client interface, service delivery system and technological options. Service innovation should take a holistic approach and its drivers are a scalable business model, comprehensive customer experience management, investment in employee performance, continuous operational innovation, brand differentiation, an innovation champion, a superior customer benefit, affordability and continuous strategic innovation [273].

The discussion on services innovation highlights that a smart city to be successful will need to develop their tools taking into account not only technology elements but also the needs of citizens and service providers, financial and legal arrangements. Chesbrough and Rosenbloom [274] suggest that business models support the creation of value through service innovation by understanding how businesses capture customer needs, organizational and financial arrangements and use the support of technology to achieve their goals.

Business models offer a solution to deal with services innovation, and innovations of all kinds have been combined with business model thinking to renew and extend common innovation and competitive strategies (e.g. [257–259]), and addressing intra- and inter-firm issues such as organizational change, value network design and innovation management (e.g. [260–263]). Its main purpose, from a practical perspective, is to allow organizations to create, deliver and capture value.

There are many business model definitions in the literature and the lack of definition creates a source of confusion inhibiting the convergence of research progress on business models. At a general level, the business model has been defined as a logic [274], a representation [262], an architecture [270,275,276], a conceptual model [259,277] and a structural template [278]. Some authors have used the concept to describe co-ordination mechanisms in economic processes [279–281] and in e-businesses [276,282]. In addition, business models are more and more related to strategic choices companies are making [283,284]. Instead of formulating business strategies, many businesses have opted to formulate business models [283]. To a large extent, a business model is determined by the concrete operational implementation of business strategy [270]. The business model is given shape by answering questions with regard to customer needs, services provided, the necessary technical, financial and human resources and capabilities and the way processes are defined, among others.

Although there is no universal definition to the concept of business models, Hawkins [285] argues that "the business model seemed to fill a niche even if no one could explain exactly what it was". Table 5.1 summarizes some of the recent definitions suggested for the business model, and their respective fundamental components.

There are a few elements that commonly turn up in definitions of business models [288]:

- **Mission**. Determining the overall vision, strategic objectives and value proposition, as well as the basic features of a product or service.
- **Structure**. This has to do with the actors and the role they play within a specific business environment (a value chain or web), the specific market segments, customers and products.
- **Process**. The concrete translation of the mission and the structure of the business model into more operational terms.
- **Revenues**. The investments needed in the medium and long term, cost structures and revenues.

Tapscott and Ticoll [281] do not directly define business models, but what they call b-webs (business webs). A b-web is a business on the Internet and represents a distinct system of suppliers, distributors, commerce service providers, infrastructure providers and customers that use the Internet for their primary business communication and transactions. Similarly, another highly network-centred approach is provided by Amit and Zott [278]. They describe a business model as the architectural configuration of the components of transactions designed to exploit business opportunities. Their framework depicts the ways in which transactions are enabled by a network of firms, suppliers, complementors and customers.

The rapid development and adoption of new technologies have facilitated organizational transformations (e.g. see [276,281]) and have increased the possibilities for the design of new boundary-spanning organizational forms [289,290] and development of new ways to create unconventional exchange mechanisms and transaction architectures [278] to deliver value. It gave rise to a new concept of business models, namely, e-business models.

5.2.1 e-Business models

Information and communication technology, that is, Internet and mobile technologies, play an increasingly important role, not only in the organizational processes (back office) but on the lives of people in modern societies. Technological adoption

Table 5.1 Selected business models definitions

Authors	Definition	Fundamental components
Timmers [276]	*"An architecture of the product, service and information flows".*	Description of the various business actors and their roles, the potential benefits for the various business actors and the sources of revenues.
Chesbrough and Rosenbloom [274]	*"The heuristic logic that connects technical potential with the realization of economic value".*	Value proposition, market segment, value chain (internal), cost structure and profit potential, value network (external) and competitive strategy.
Osterwalder et al. [277]	*"A business model is nothing else than a description of the value a company offers to one or several segments of customers and the architecture of the firm and its network of partners for creating, marketing and delivering this value and relationship capital, in order to generate profitable and robust revenue streams".*	Product innovation and value proposition, customer management, infrastructure management, the capabilities and resources, value configuration, network, partnerships and financial aspects.
Morris et al. [262]	*"A concise representation of how an inter-related set of decision variables in the areas of venture strategy; architecture and economics are addressed to create a sustainable competitive advantage in defined markets".*	Value proposition, customer, internal processes, external positioning, economic model and personal/investor factors.
Johnson and Kagermann [286]	*"A Business model consists of four interlocking elements, that, taken together; create and deliver value".*	Customer value proposition, profit formula, key resources and key processes.
Janssen et al. [287]	*"A business model reflects the core business of an organization and is useful to describe (and even prescribe) the organization from the perspective of its main mission, and the products and services that it provides to its customers".*	Business logic, value proposition, customers, current or future business.

and developments have opened new horizons for the design of business models by enabling organizations to change the way they interact with customers and suppliers, and organize and engage in economic exchanges both within and across firm and industry boundaries [291,292]. The Internet is a principal driver of the surge of the interest for business models and the consequent emergence of a literature which revolves around the topic (e.g. see [293–295]). As a consequence, the research stream which has devoted the greatest attention to business models is that of **e-business**. e-Business refers to *providing digital services* and *conducting commercial transactions* over the Internet (see also [280]). It comprises *Internet-based business, e-commerce, e-markets* and *Internet-based business*. A study conducted by Shafer *et al.* [296] reviews that out of 12 definitions in established publications during the period 1998–2000, 8 were related to e-business. Scholars have accentuated different aspects of new business models – from the ways companies exploit supply chain reconfiguration (e.g. value chain dis-intermediation or re-integration) to the ways revenues are collected (subscription cost and fees from the customer, advertising and sponsoring revenue from other firms, commission and transaction fees from provided services, etc.).

Several scholars have attempted to classify e-business models by describing types. The attention to the classification of business models was driven by the emergence of the Internet and the new opportunities for companies to conduct business electronically [297]. The use of taxonomies made it become clear in what respect the Internet business models would be different from business as usual. Functional integration and the degree of innovation were two key dimensions on which Timmers [276] based his taxonomy of 11 business models: e-shop, e-procurement, e-auction, e-mall, third-party marketplace, virtual communities, value-chain service provider, value-chain integrators, collaboration platforms, information brokerage, trust and other services.

Weill and Vitale's [298] classification of e-business models encompassed eight *atomic* business models: content providers (information, digital products and services) via intermediaries, direct to customer providers, full-service providers (e.g. financial, health) directly via allies, intermediary who brings together buyers and sellers by concentrating information, shared infrastructure who bring together multiple competitors to co-operate by sharing common backbone infrastructure, value net integrators (gathering, synthesizing and distributing information), virtual community and whole-of-enterprise/government (consolidation of all services). Applegate [299] proposes a taxonomy of six business models: focused distributors, portals, producers, infrastructure distributors, infrastructure portals and infrastructure producers.

The commonality of these taxonomies is that they all attempt to describe and organize business models archetypes around typologies and taxonomies which are mainly enabled by Internet technologies. On the other hand, Dubosson and Torbay [275] propose a multi-dimensional classification scheme for business models. They identify the following principal dimensions for classifying business models: user's role, interaction pattern (provision of services), nature of the offering (information, services or products), pricing system, level of customization (mass vs. customized content) and economic control (from self-organizing to hierarchical). Besides providing

Table 5.2   e-Business models components

Authors	Fundamental components
Mahadevan [280]	Value stream, revenue, logistical stream
Afuah and Tucci [297]	System architecture, customer value and revenue sources
Osterwalder *et al.* [277]	Value proposition, customer segments, partners' network, delivery channel and revenue stream
Brousseau [300]	Costs, revenue stream, sustainable income generation, goods and services production and exchanges

typologies that enlist and describe various generic e-business models, scholars have also investigated the fundamental components of e-business models. Table 5.2 presents a summary of these efforts.

Government organizations are challenged to provide more customer-oriented products and services and to approach customers pro-actively. For this purpose, governments are increasingly exploring different types of web-based business models to serve customers through multiple channels such as web-based services to deliver smart cities solutions. In government, rather than seeking monetary profit the business models of government digital strategies aim at using the Internet to add value to the constituents from delivery to political involvement [287].

While there exist opportunities for governmental organizations to transform their current practices and provide new products and services which drives innovation and sustainability initiatives, many initiatives as discussed in Part I of this book remain at the basic Internet presence stage (digital cities), the types of transactional services available remain very limited, and the ideas developed to this point remain abstract. In order to fully experiment and improve services – and thus their business models – it is necessary to comprehend the basic elements or components of business models. However, there is no clear consensus of what constitutes business model for government digital strategies, and thus no established general business models framework. Based on a study of 59 e-government websites, Janssen *et al.* [287] classified eight e-government business models, which are based on the eight atomic business models [298], but adapted for e-government. These business models are

1. Content provider
2. Direct-to-customer
3. Value-net-integrators
4. Full-service provider
5. Infrastructure service provider
6. Market
7. Collaboration
8. Virtual communities

Janssen *et al.* [287] found a number of elements that are present in such e-government business models, and they are:

- Derived from the main mission of the public organization, often founded in law;
- Contains the logic and elements to fulfil the mission successfully using the Internet, and to satisfy citizens and/or businesses;
- Describes the products, services and mix of channels;
- Addresses the relationship between an agency's strategy and information systems;
- Describes the position in the organizational network and relationships with other agencies that target the same audiences;
- Describes future evolvement;
- Ideally independent of temporary technology.

Thus far the quest towards new government business models for smart cities remains problematic. There is a need to develop and apply business models to continue the progress towards creating smart cities and accomplish citizens orientation. Yet the exact link between e-business models and smart cities initiatives is yet to be explored. While high-quality experiences with responsive, integrated, web-based services in the private sector have led citizens to expect the same from the public bodies and agencies [301], how governments can harness web-based business models to improve their digital strategies that will enable the creation of smart cities remains relatively unexplored in the literature. However, some authors have begun to deal with business models applied to related strategies, such as e-government business models components [287], e-government initiatives [302] and mobile services [270].

In addition to classification of business models, some frameworks for more detailed analysis of e-government business models can be found in the academic literature. Bakry has defined a STOPE model for e-government initiatives. It consists of five domains for e-government application business modelling, namely, Strategy, Technology, Organizations, People and Environment [302]. Partially based on the STOPE model and building further, Esteves and Joseph's construct EAM (e-government assessment framework) a three-dimensional framework for the assessment of e-government initiatives, based on maturity level, stakeholders and STOPE domains [303].

Ballon [304] proposes a holistic business modelling framework called *Business Model Matrix* that is centred around value network, functional architecture, financial model and the value proposition parameters that describe the product or service that is being offered to end users. However, its business models are relevant to the closed systems approach in which the public component is outside the value network. Walravens [305] extended the Business Model Matrix to support mobile services in cities and have governance and public value as two fundamental elements. Based on business models and business model frameworks previously developed, Bouwman and Haaker [270] introduced a holistic model for describing the business models of electronic services, called the STOF model. STOF hides the complexity of many other models into four core components or domains: namely Services, Technology, Organization and Finance. The STOF model was developed for mobile services and also provides a means to analyse and develop the business models of a service over time, as the service develops from an initial concept to a service in the market. Unlike some other business model frameworks such as Osterwalder's business model canvas [277], the STOF method takes into account techno-economic interdependencies, which are the key aspects of the smart cities platform research.

However, all these models remain at a high level and provide little or no help in the actual service design process of government digital strategies, and therefore their application to smart cities is limited. While the strategy and innovation mainstream treats the business model mainly as a mediator between technologies, strategies and economic value (e.g. [258,259,274,286,306]), the question of how business models can support smart cities and their stakeholders, and their innovations in creating, delivering and capturing economic, social and sustainability value has so far received little attention. While there are many definitions for business models and many different components, existing business models can be summarized in terms of strategic choices, value creation, value networks and capturing of value [296]. The existence of so many different classifications of components illustrate the lack of a common framework.

The next chapters of this book seek to contribute to this new scholarly field by exploring the theoretical inter-relations between smart cities and business models that offer value to users and allows the creation of value through sustainability innovations and new business models through integrated data. We argue that the interdependence between city data infrastructures and business models takes two forms. First, city data offering needs business models to enable the normal course of its supply chain operation. Second, business models are influenced by external forces, which ultimately impacts the offering of city data. Furthermore, a business models framework must be dynamic and adaptive during different phases of the development of the data infrastructure strategies of cities and also to respond to any changes in market, regulations and policies. Moreover, as cities specialize in their core competencies – policymaking – a large number of stakeholders must be organized in a value network of partners in order to provide the necessary services and technology to support the smart cities development. To the best of our knowledge, however, existing research focuses on single-provider platforms and no research has been developed on common data service platforms provided by multiple actors in smart cities.

To put these principles into practice, our business models framework combines city data offering with a business model thinking to renew and extend common innovation and competitive strategies [257–259] and address intra- and inter-firm issues such as organizational change, value network design and innovation management [260–263]. From a practical perspective, the main purpose of our framework is to allow governments to create, deliver and capture value through data infrastructures which are designed on the basis of social influence and not authority.

5.3 The framework for data infrastructure design

Part I of this book highlighted the need for methods to enable data as infrastructure to realize smart cities. This work aims to address the need. The hypothesis of this work, stated in Chapter 1, is that dynamic business models and supply chain techniques provide effective support for city data management and orchestration. Enabling data as infrastructure in smart cities requires the investigation of several issues from different disciplines. As we have discussed in Chapter 2, on the one side, the management of city data involves a range of technical issues, such as bringing technology and non-technology components of the city together to explore how a smart cities initiative fits into the wider city context, understanding the semantic challenges introduced by existing data heterogeneity and the non-existence of a common data model, eliciting and modelling requirements to design the smart city strategy.

On the other side, it also encompasses a number of non-technical issues, for instance, managing the conflicting expectations of the stakeholders of smart cities with regard to the smart city, establishing and designing business models that strike a balance between collaboration and competition. In order to improve the quality of decisions made during the design of data as infrastructure strategies, policymakers should be provided with processes and strategies to handling technical and non-technical concerns that may arise during the design of smart cities. In this book, we aim to address both classes of issues.

An alternative to the prevailing top-down approaches adopted in previous studies, this book presents a *middle-out* approach to model such an ultra-large and highly interconnected platform. Platforms and the ecosystems that grow up around them are acknowledged as *drive innovation and transform industries* [61]. They form the context for both co-operation and competition, in ways that potentially combine the best of both [307]. Open platforms can orchestrate people, technology and organizations into socio-technical ecosystems that remain, even on large scale [308], flexible and innovative. For data infrastructures, we argue that socio-technical ecosystems are the only mechanism that will allow sufficient resources to be brought together in a co-ordinated way to produce the systems we need.

In the context of data infrastructures, we recommend that cities integrate the data infrastructure supporting activities that are citizen-oriented services, technology, value network and governance into our framework, named SMARTify, to orchestrate the primary activities of the city data supply chain and guide their business models. This business model framework is used to capture the requirements of the platform. The business model framework is used alongside the Stakeholders Dependency Diagram of the KAOS goal modelling framework [309]. Ideally, the business models framework should:

- Model digital infrastructure business models and capture requirements from a *global* perspective.
- Be independent of the individual doing the analysis.
- Not overload decision makers with information.

The development process from business models to established data infrastructure can be divided into a number of phases: Research and Development, Procurement, Roll-out and Market. It enables the value proposition of the data infrastructure to be studied over its entire life cycle. To validate the business models framework, we use critical design issues (CDI) and critical success factors (CSF) that provide support for the development of the urban capabilities described in Chapter 2 (Physical, Human, Social, Institutional, Innovation, Sustainable, Economic). CSFs are based on CDIs and iteratively refined to create a viable business model.

The second part of the SMARTify approach is to investigate the logistical distribution of city data. The physical function of the city data supply chain is very important for the provision of efficient city services. By extending the simple reasoning that the physical flow inputs and outputs traverse the supply chain for a broader view of the value chain, we realize that it is necessary to take into account the relationships that make this chain more competitive with lower costs and lower risks for each of its members. In order to manage and co-ordinate a networked movement of information from data/metadata suppliers to data consumers, this research addresses the problem of city data management under the aegis of a closed-loop supply chain. Closed-loop supply chain or reverse logistics are designed and managed to explicitly consider activities along both the forward and the reverse flow chains [310]. Reverse logistics in closed-loop supply chains play a central role in improving economic benefits, as well as in achieving the desirable goals of an *eco-efficient* data ecosystem (e.g. data re-use, feedback loop).

The functional and non-functional requirements gathered from the business model framework serve as an input to a closed-loop supply chain model, which is designed and managed to explicitly consider all the activities and processes that create competitive advantage along both the forward and the reverse flow of the data production chain. The data value chain proposed in this book consists of all distinct data activities inside the chain that starts with the gathering of raw data, passing through data operations to context enrich data and transform data into information, finishing on the integrated and rich content knowledge to be delivered to the stakeholders. The closed-loop framework can be represented as an initial architecture view of a smart city open data framework.

The SMARTify approach models and guides the stages of definition of high-level requirements and assumptions, as well as determining structural preferences and a reference architecture. The design method explicitly addresses questions regarding the four domains and takes into account CDIs as well as CSFs related to supporting the urban capabilities of smart cities. Figure 5.1 contains an overview of the various steps in the design process that was used in the case studies. Although the figure presented above suggests that designing a business model consists of consecutive phases, it is actually a continuous iterative process because SMARTify is a dynamic business models framework.

The results from evaluations provide evidence to support the hypothesis that business models and supply chain management techniques can provide effective support for designing data infrastructures for smart cities.

5.3.1 Step 1: Business models outline

The first step in the SMARTify approach is the application of the business models framework and the analysis of the components of all domains. Figure 5.2 illustrates

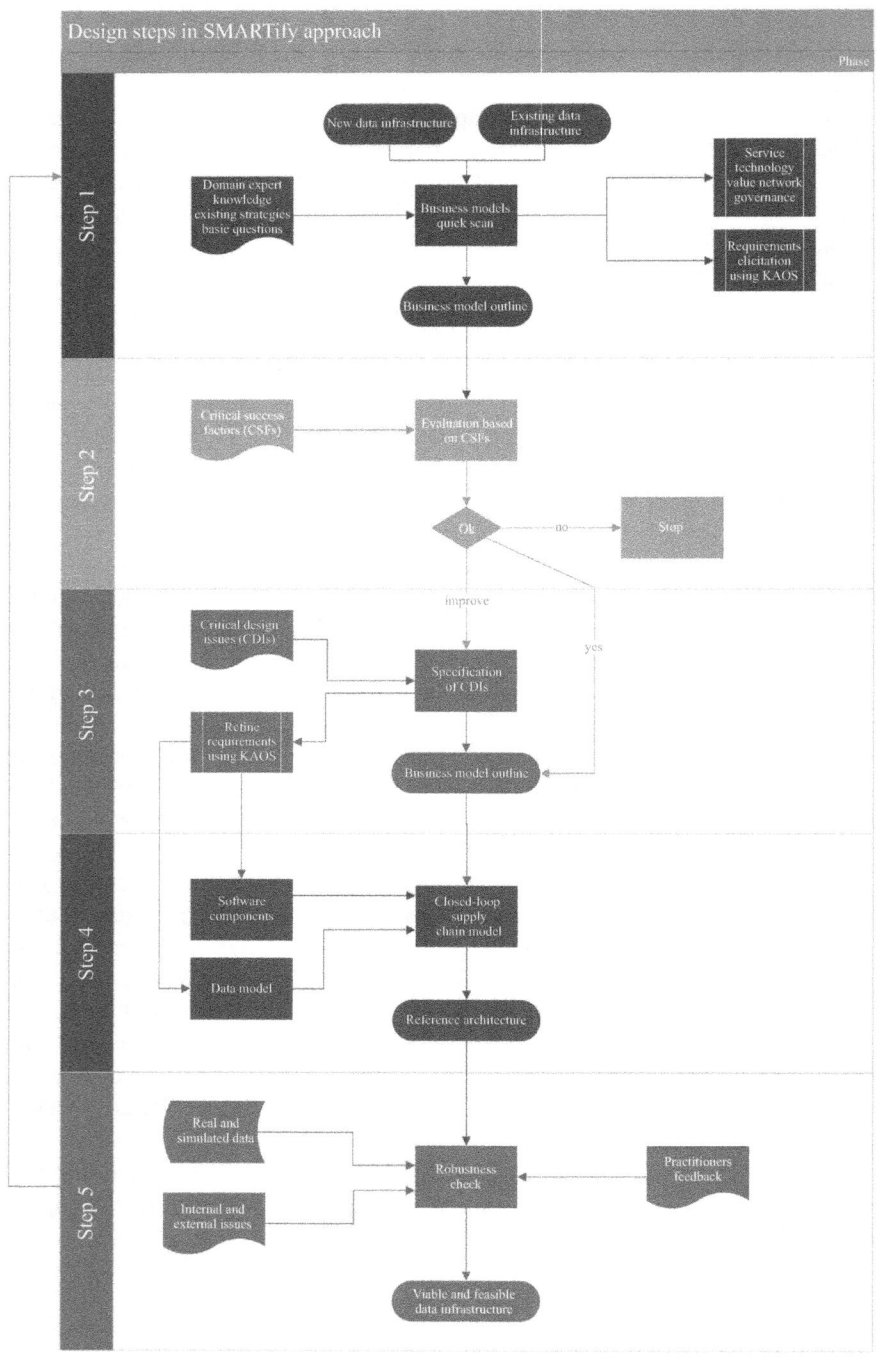

Figure 5.1   Design steps in SMARTify approach (adapted from [270])

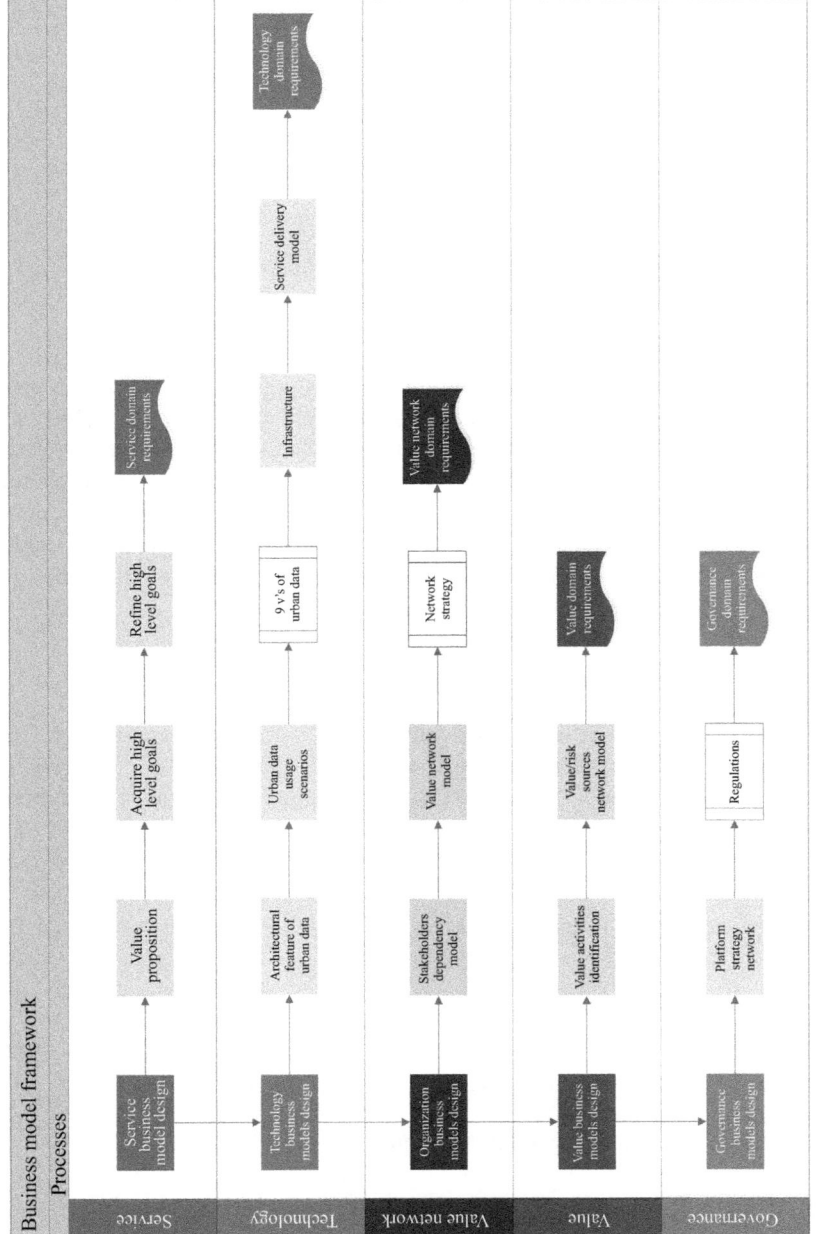

Figure 5.2 Business models framework components

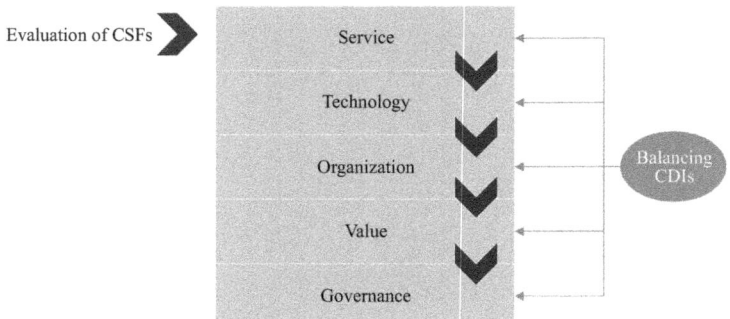

Figure 5.3   Specification of CDIs per domain (adapted from [270])

the components of the business models framework proposed in this book. To provide answers and build the business models outline, one can cluster the questions based on potential data sources, such as consumers, the providers or specific partners. Next, one needs to formulate specific tasks and indicate how the necessary answers will be generated, for example, through interviews, market research or desk research. The answers are formulated in deliverables which eventually provide insight into possible ways of formulating the business model as well as the critical design issues.

5.3.2   Step 2: Evaluation with CSFs

The evaluation of the Quick Scan design looks at how well the business model satisfies the CSFs for smart cities capabilities. The evaluation in Step 2 focuses on the CSFs. The underlying logic is that a negative assessment of certain CSFs implies that there will be bottlenecks in the business model's viability and that CDIs related to such CSFs should be redesigned [270]. The evaluation is based on the causal models presented in the next section, which state that design choices with regard to the CDIs have a strong influence on the CSFs, and thereby on business model viability as well.

5.3.3   Step 3: Specification of CDIs

Depending on the evaluation of the CSFs in Step 2, a selected set of CDIs is specified in greater detail. There are two ways to approach this refinement:

1.   Refine CDIs for each of the domains (Figure 5.3)
2.   Refine sets of CDIs for each of the CSFs (Figure 5.4)

In addition, performing requirements trade-offs are also described, that is, conflicts within the relationships of requirements from different business model domains, and the criteria that guide the selection of a specific requirement over another are explained in Part IV of this book.

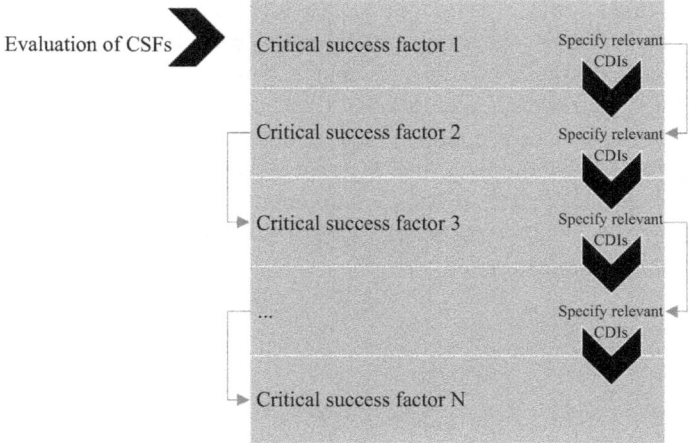

Figure 5.4   Specification of CDIs per CSF (adapted from [270])

5.3.4 Step 4: Closed-loop value-chain model

The data value chain presented in this book consists of all distinct data activities inside the chain that starts on the gathering of raw data, passing through data operations to context enrich data and transform data into information, finishing on the integrated and rich content knowledge to be delivered to the stakeholders. In Step 4, based on the requirements gathered in the business models we identify the software components, technical functionalities and the soft goals of the data infrastructure. This reference architecture can serve as a blueprint of the infrastructure to be built or an input to a software simulation experiment. The value-chain model is presented in Chapter 7 of this book.

5.3.5 Step 5: Robustness check

Step 5 involves the evaluation with respect to robustness and adaptivity. Business models evolve over time under the influence of the external business environment, and the robustness of the business model has to do with the ability to cope with changes in the smart cities environment. In addition to robustness, the business model's capacity to adapt to external influences is an important evaluation criterion. Typical examples of external influences are changes in user requirements, regulatory changes and accommodation of a new technology. This stage involves evaluating the design with experts through interviews and questions and/or through software simulation when necessary. For instance, having scalability as a critical design issue makes necessary to guarantee the data infrastructure is able to scale with regard to users, data and services.

    The remaining chapters of Part II describe each part of the business models framework and the reference architecture for data infrastructures based on closed-loop value-chain models (Part III). We believe that our comprehensive framework for data infrastructures design can provide guidance for policymakers wishing to grow the network of users, vendors and service providers that surround platforms and can also begin to provide more general design principles and move us toward a more systematic understanding of platforms. This is the first time such an extensive investigation and complete approach has been applied to design a data infrastructure for smart cities which is supported by multiple actors. We expect the content of Part II of this book may fill this knowledge gap and assist practitioners and policymakers on the design of platforms in the context of smart cities, as well as offer opportunities to re-think an integrated urban infrastructure.

Chapter 6
The business models framework

6.1 Introduction

Understanding of smart cities and their business models are important in studying how one can best provide value to its stakeholders. Our approach provided through a data infrastructure proposes to improve innovation, development, user engagement and stakeholders' collaboration of smart cities services. Service innovation is directly related to the business models that support these services [270].

As discussed in previous chapters, establishing common service platforms for smart cities' services requires data, resources and expertise across disparate sectors of telecommunications, transport, energy and health care and so forth. Despite considerable research being devoted to design applications for smart cities, less attention has been paid on the actors who will interact, use and promote such solutions. The stakeholders and policymakers who may be involved in the development of a platform for city data management have different backgrounds, interests and expectations, as well as several potential sources of conflicts. Hence, the first and foremost organizational issue is how collaboration for establishing common service platforms for smart cities' data services may arise.

It is important to understand the motivation and criteria which organizations take into account when deciding to join a collaborative project in establishing a common service platform. Therefore, equally important is to strike a balance between collaboration and competition [307] and build up trust and commitment between those parties to maintain collaboration and deal with power struggles [311–313]. The provision of cross-domain city data can be facilitated through the development of an intelligent data infrastructure based on a middle-out leadership pattern. Rather than acting solely as *implementers of initiatives*, our approach acknowledges the importance of the expectations and needs of the users or data infrastructure (e.g. citizens, businesses, entrepreneurs, data scientists, among others). It takes social influence into account while maximizing the efforts of other stakeholders who are working towards the achievement of the same goal: to create better data infrastructure which will unlock the data that realize smart cities.

We argue that a one-size-fits-all approach to city data management transformation and simplistic approaches to engage stakeholders of city data are unlikely to work. Rather, we focus on the enabling processes and activities by which innovative use of technology and data in the context use of a city's inhabitants, alongside governance strategies supported by a strong value network of partners, can help deliver the various visions of data strategies for cities in more efficient, aligned and effective ways. Embracing the technology and non-technology components of data infrastructures presented in this book will ensure that standards are adhered to, interoperability is guaranteed, smart governance is in place, a strong value network of partners are built, and feedback is facilitated. The use of our framework will help transform current data management practices and facilitate the creation of a data marketplace within both the public and private sectors facilitating the exploitation of city data in smart cities. This implies a focus on providing a business models framework for cities to:

- establish a clear, compelling data management approach which establishes collaboration with data owners across the city;
- facilitate the exploitation of rich, interoperable and engaging cross-domain city data to enable deep knowledge discovery, and the creation of new valuable integrated services;
- take a citizen-centric approach to all aspects of data provision;
- enable a ubiquitous, integrated and inclusive real-time offering of city data on an open and interoperable way;
- offer smart data to boost stakeholder-led innovation by communities, interested citizens and all city sectors;
- embed legal agreements and policies to enable a thriving market in the re-use of both private and public data together in appropriate ways;
- build partnerships to deliver holistic and interoperable solutions;
- offer mechanisms for stakeholder's feedback on service provision, accessibility and how users are extracting value from city data;
- enable new profitable business models to emerge, and the creation of a wide range of new and engaging services in the city.

Figure 6.1 illustrates a holistic overview of our framework. The first stage of the theory development of the business models framework consists of collecting empirical data concerning multiple examples of contested approaches to city data management. In this stage, we first introduce an example that illustrates the class of data services our framework model aims at supporting; we define what data, technology, stakeholders and context are relevant to these services. This example is expressed as stories involving the value proposition of the platform providers, the complex sequences of interactions between platform providers and other stakeholders as they conflict over the requirements of the platform, data ownership and services innovation.

The examples used in this book come from the literature on smart cities, open data and city data management which are rich in narrative detail regarding examples of these tensions. Stories found within the dataset are analysed in order to produce a simple classification of the different actions employed by all or some stakeholders attempting to implement innovation. This classification is applied across the different stories concerning contested approaches to city data management in order to produce an abstracted set of patterned sequences of actions describing a tension and a

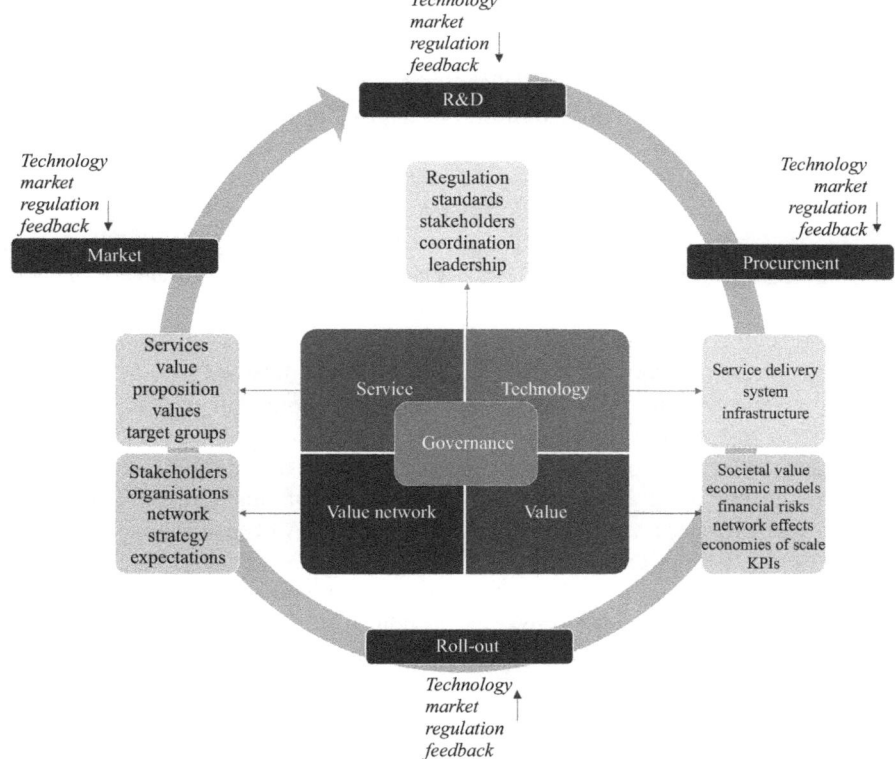

Figure 6.1   SMARTify – the business models framework

disregard for the nature of city data. In particular, we focused on the many levels of complexity raised by business models, value networks, feedback loops, standards, regulations and licensing models, as well as efficient data governance arrangements that a data infrastructure for large-scale city data management would need to support. The creation of the business models strategy framework was built upon the extensive knowledge available in the literature on general business models, especially [270], and we extended and adapted the model to serve in the context of large-scale data infrastructure design.

The framework is divided into five broad domains

- **Service domain** focuses on the delivery of integrated value propositions. This domain encompasses tailor-made data services which carefully target the needs of users and businesses and explore use cases where data is used to deliver different forms of value.

- **Technology domain** enables the widespread exploitation of data technologies which act as an enabler for the proposed services. This domain focuses on creating agreements between stakeholders regarding data handling, standards and technical infrastructure to allow the design of infrastructures that will serve as the foundation for widespread exploitation of data in the long term.
- **Value network domain** establishes a value network to maximize the efforts of the stakeholders of smart cities. The stakeholders of data infrastructures must be brought together to solve the current technical, strategic, organizational and regulatory issues in the city data supply chain in order to enable users to exploit new business models. This domain covers the roles of the value chain which city leaders can use to steer a strong value network of collaborators who will provide the expertise needed to deliver a data infrastructure.
- **Value (societal and financial) domain** explores new and innovative business models for value creation. This domain includes financial values which details prices, transformational business models non-monetary values created by new services that are created by increased access to data and closer integration between city systems. Cities are given the clarity they need to change existing processes in order to capitalize on these.
- **Governance domain** establishes governance arrangements to create impact and accelerate smart cities growth. This domain comprises collaboration and leadership models to articulate, measure, manage, deliver and evaluate the data infrastructure in practice. It covers key aspects to guide cities on how to address city-wide challenges of joining-up across city silos and orchestrating the data marketplace.

The main component of the service domain is the users' **value** of a product or service. As the customer value of the service is the most relevant aspect of the service, it serves as the reference when comparing to the other domains. Although technology is basically a driver for new innovative services and business models (push model), from a customer perspective technology is only an enabler. In the latter case, technology pull plays a central role, one that can only be understood from a customer value perspective and one that requires an understanding and elaboration of user requirements. After discussing the service and technology domain, we address the organization domain, the way resources are made available, and the value domain, which includes investments as well as, for instance, pricing strategies, increase in transparency and innovation in the city, and lastly the governance domain which includes the platform leadership model strategies, as well as regulations.

Each domain description starts with theoretical notions about the relevant concepts and issues with regard to that domain. The concepts that are most relevant from a design perspective are subsequently addressed and included in a descriptive conceptual domain model. The relationships between the concepts within and between the domains are discussed. The five descriptive domain models together provide a descriptive conceptual framework for the design of business models.

In the following sections, we discuss these five domains in greater detail and also take a closer look at the theoretical and technological concepts that are the basis for our framework and that will lead the design of business models.

6.2   Service design

This section will be concerned with identifying the key service features of smart cities that determine the strength of the services provision and hence user's take up. The goal of services design for a city in the smart city arena is to find a position where the city can best position itself in delivering competitive urban services and can influence them in its favour such as in attracting new businesses, new residents, empowering its ecosystem of new businesses and scale-ups. Since the strength of services offered in smart cities is a result of meeting user's expectations and needs, the key for developing a strong service design is to delve below the surface and analyse the elements that influence how successful a service is. In the context of services design, smart cities seek to deliver new digital services that will address the city and societal needs of cities in a positive manner. Therefore, data infrastructures services design must be a process that both city and citizens are active co-creators of such services. Knowledge of the underlying elements unlocking the *user's value of a service* highlights the critical strengths and weaknesses of the city's digital technology, value sources and activities; clarifies the areas where strategic changes may yield the greatest user's impact; and highlights the areas where technology and value network of partners promise to hold the greatest significance as to solve urban challenges and needs.

The concepts of the service domain and their relations are mapped in Figure 6.2. In this figure, the grey boxes within the container represent the concepts of the service design and their relationships, whereas the outside boxes represent concepts from the other domains (technology, organization, value and governance) which directly impacts and/or influences the concepts of the service design.

In this figure, the central components of the service domain are the user **value** of a product (e.g. data) or service. As the customer value is the most relevant aspect of the service design, it serves as the reference when comparing and benchmarking different options. The user value can be seen as an improved or new, innovative offer by the city to its residents. Chen and Dubinsky [314] describe *value* as part of an equation in which customers in target markets compare the perceived benefits and total costs and/or efforts of (obtaining) ownership of a product or service. Kotler [315] emphasizes that *human needs and wants* must be the starting point for service offering:

> "a human need is a state of felt deprivation of some basic satisfaction, wants are desires for specific satisfiers of these deeper needs and demands are wants for specific products that are backed up by an ability and willingness to buy them".

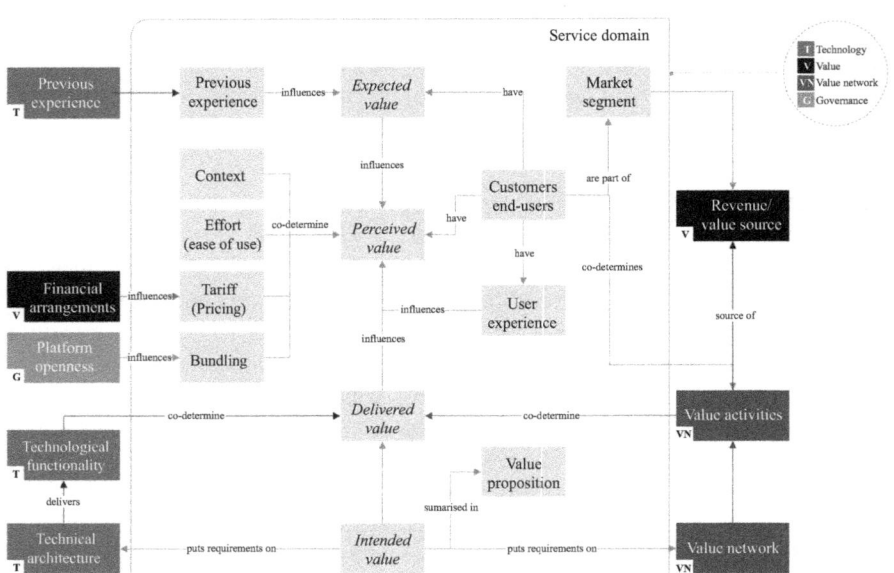

Figure 6.2   The service design (adapted and extended from [271])

This idea has been validated in other research works which also highlight that in our current modern society, the products and services must both satisfy needs and provide valuable customer experience [294,316]. As such, the **value proposition** of a smart city – which is delivered through electronic channels – must be recognized as being better, and as outperforming competitions with regard to human needs and experience.

The concept of **value** can be broken down into four sub concepts: **expected value, delivered value**, **intended value** and **perceived value** [270]. The human needs and previous experience with previous versions of the service or with similar or alternative services influence the **expected value** a customer has on a service. For instance, disabled residents expect the city to have wheelchair accessibility on its entire transport network and streets so that they can exercise their *right to the city*. Unlike Washington, DC, in which its entire transport network is wheel-chair accessible, most cities in the world only have a few stations which are wheelchair accessible (the delivered value). The problem is further exacerbated when disabled residents are unable to obtain accurate information on the location of all accessible stations or streets – an example of a poor perceived value. Very often due to all kinds of organizational, technical and operational constraints the intended customer value is not the value that will be ultimately delivered to the customer. The delivered value of a service is affected by technological functionality and the value activities of the platform. In some cases, the intended value and the delivered value may be the same, but even in those cases, it is not always the value that will be perceived by the user [270]. In our example, if the city

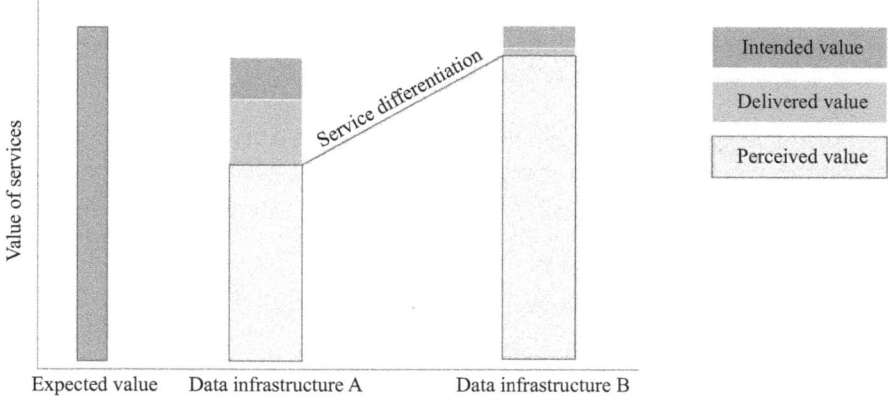

Figure 6.3   Concept of value and service differentiation

consider offering ready-access to accessible stations data across its transport network, firms can deliver applications that enable users to create routes that are accessible to wheelchair users. This intended value is translated into functional requirements for the technology architecture to make the data available and into requirements for the formation of the value network (e.g. partners to collect data, develop API's for data access, etc.). This simple improvement in information access can provide a significant impact on the liveability of residents considering that approximately 131,800,000 or 1.85% of the world's population require a wheelchair.[1]

The **perceived value** is the value a customer perceives when using a particular service and it can be defined as the difference between Delivered Value and Expected Value. Perceived value is user-specific and can only partially be controlled by the platform provider. Thus, the value is subjective to the user, and the provider of the service will not be able to serve all users with the same experience of value. In many cases, customer value as perceived by the end-user has little to do with that which that is envisaged in initial business models, depending to a large extent on personal or usage context [314]. Figure 6.3 illustrates the concept of value and its impact on service differentiation between data infrastructure provider.

The **user experience** while interacting with services (e.g. usability, feeling of security and trust) is beyond the control of the service provider and it affects both perceived value and the value activities. A **service innovation** is only successful when it provides benefits to the customer in a particular **context** [314]. Context is crucial to successful innovation and serves as the starting point in the formulation of the Intended value of the service domain. The demand for city data can shed the lights on the possible expected value that the customer and end-users expect from a

---

[1] https://www.wheelchairfoundation.org/programs/from-the-heart-schools-program/materials-and-supplies/analysis-of-wheelchair-need/

data infrastructure. Here, we outline five distinct processes of city data use defined in [317] and extended here. Note that they are not mutually exclusive and many users of city data employ multiple usage patterns.

- **Data to fact:** often individuals may seek out specific facts in a newly open dataset. These facts may support their engagement in civic or bureaucratic processes, or in business planning, or may inform personal choices. Facts could be found through online interfaces but also by browsing downloaded csv documents and spreadsheets.
- **Data to information:** creating a static representation and interpretation of one or more data sources, information can take the form of visualizations, blog posts, info-graphics or written reports.
- **Data to interface:** creating a means to interactively access and explore one or more datasets. For example, creating a searchable mapping mash-up, or providing a tool to browse a large dataset and crowd-source feedback or scrutiny. Interfaces often also include static interpretations of data (data to information), showing particular summary statistics or algorithmically derived assessments of underlying data.
- **Data to data:** sharing derived data (either simply an original dataset in a new format, or data that is augmented, combined with other data, or manipulated in some way). A whole dataset may be shared; an Application Programming Interface (API) onto a dataset created; or an interface that makes it easy to download subsets of a large dataset. API may also enable the sending of data updates to users in real time, avoiding the need to re-download entire files or to search for new entries.
- **Data to service:** services where Open Government Data plays a – behind the scenes – role either online or offline. For example, the use of boundary data to route messages reporting potholes to the responsible transport authority.
- **Data to machine:** data collected or input into machines in the urban environment. For example, the collection of air pollution data from sensors embedded in trains and in air quality stations. This data can be processed and sent to other systems or machines such as variable message signs containing warning messages to citizens in case of unsafe air quality levels.

In mobile services, a customer is considered as the individual paying for the service and end-user as the individual actually using the service [270]. The customer and end-user roles coincide in regards to consumer services; however, they differ to business services and are considered different entities. In the consumer services provided by data infrastructures, we define *all stakeholders using the resources of the data infrastructure* as **customers/end-users** (similar role), and in the business services scenario we define *stakeholders providing data (open/proprietary) or services* as the **customers of the data infrastructure** since they provide the products and services to be offered.

Both customers and end-users make an **effort** to use the service (e.g. learn how to use the service, the effort it takes to find a particular data). **Effort** refers to all non-financial effort an end-user/consumer must make to efficiently use the platform. Existing Diffusion of Innovation literature suggests that innovations that are perceived by individuals as having greater relative advantage, compatibility, trialability, observability and less complexity will be adopted more rapidly [318,319].

Aligned with previous research, we highlight five key factors which influence the take up of a service innovation in smart cities, and they are outlined below. Out of the factors, three are related to the effort concept. The lesser the effort one has to put in order to use a service the easier is to use it. Effort and tariff influences the perceived value [320], and therefore, it affects the success of a service that can be measured in terms of adoption and usage. In the next sections, we elaborate on the first two factors, and the last two are widely covered in many chapters of this book.

1. The comparative advantage of the service innovation over other existing services;
2. Service's neutrality, fairness and accountability;
3. Complexity and testability of the innovation;
4. Compatibility with the city context (e.g. demographics, culture).

6.2.1 Comparative advantage of the service innovation

The comparative advantage of the innovation over other existing services refers to the service differentiation. Services differentiation in smart cities involves the provision of unique services that possess high perceived value by their users. Porter's research on competitive advantage [321] highlights that differentiation can be achieved at any stages of the value chain and serve the individual or a group of users with particular needs. As such, any activity along the value chain of city data value chain is a potential source of uniqueness. We explain the city data value chain in more detail in Part III of this book.

The concept of individualization makes assessing needs and careful targeting crucially important for realizing the potential of providing tailor-made data services solutions. Market segmentation involves dividing the market (of e.g. consumers) into groups with shared needs or desires, requiring a separate service focus [322]. In smart cities market segment can be for instance customers within government, policymakers, interested citizens, professionals/researchers within non-government or government organizations. As an example, it cannot be assumed that all consumers of data services will have the same competences in data analysis. Therefore, providing easy-to-use interfaces and accessible documentation regarding datasets and APIs can assist users in fully exploiting the value of data.

For more advanced users, providing more flexible query capabilities and data versioning may reduce the effort required from users to discovering city data. These advanced features can be a strong source of differentiation. The current adoption barriers of open data [192] are the users' frustration at the existence of too many data initiatives and lack of ability to discover the appropriate data as data is static and presented in different formats, inconstant quality, are not integrated with any other sources of data on the web, do not present a well-defined semantics/metadata.

6.2.2 *Service's neutrality, fairness and accountability*

The concept of service's neutrality refers to the availability/unavailability of digital services and/or data without the influence of political agendas or political ideology, that is, they must be available impartially. Data and services in smart cities must be considered as neutral and objective reporting the truth about the city environment. They should provide to users' segment/groups always respecting data license, regulation and privacy laws. In a similar fashion, the digital services and the backbone technology – including algorithms – should be free from ideology in their conception, operation, integration and dissemination.

The concept of service's neutrality in smart cities would enable the design of fairer, accessible, safer, more secure cities that would not become a commodity to be exploited by a singular proportion of the population. This is aligned with Harvey's [323] research on **our right to the city**:

"The right to the city is far more than the individual liberty to access urban resources: it is a right to change ourselves by changing the city. The freedom to make and remake our cities and ourselves is one of the most precious yet most neglected of our human rights".

However, as many authors have argued, data are not free from the influence of views, ideals, technologies, people and contexts that generate, process, persist, analyse and distribute them [324,325]. As such, data and services innovations in smart cities are highly susceptible to not being socially benign but the result of choices and constraints shaped by a system driven by public, political, financial, ethical or regulatory considerations and opinions. The result is that data and services become inflected by social privilege and social values.

There is no doubt that data and digital services innovation provide data and tools that are useful for managing and improving city services; however, the politics and limitations of such data and the methods used to produce and analyse them need to be scrutinized and examined as to the values and agendas underpinning them and whose interests they serve. Data and services need to be complemented with a range of other instruments, policies, ethics and practices that are sensitive to the diverse ways in which cities are formed and function. The consequences of not doing so are the exclusion of citizens from the provision of services.

As cities and governments rely even more on automation and machine learning such problems may only be detected when they become a social issue. A classic example of lack of service's neutrality in the provision of digital services in cities is that of Amazon's Same Day Delivery. A research conducted by Bloomberg demonstrated that Amazon's Same Day delivery service was not offered to many African American communities in the USA.[2] Although Amazon's claimed that racial information was not embedded in their algorithms for service's targeting within American cities,

---

[2] https://www.bloomberg.com/graphics/2016-amazon-same-day/?cmpid=google

the lack of fully examining who was serviced led to the exclusion of many potential valuable costumers. Recent research has raised questions about the data used in smart cities and their effect in revoking the right to the city from elderly and economically deprived citizens [326,327]. In Boston, for instance, less privileged neighbourhoods experienced a significant higher digital divide and were not able to report city maintenance issues through digital channels in comparison to more advantageous citizens.

Hence, cities need to be transparent about the reasoning behind the services they provide and data they collect and in what sense this data is repurposed and used to decide how services are delivered and to whom and what is the reasoning behind such decision making. Reducing the opaqueness big data analytics and intelligence can efficiency assist in ensuring services are provided to all without political motivations. Furthermore, disruptive technologies such as artificial intelligence and the Internet of things (IoT) are likely to have a major and widespread impact on the nature of jobs. On the one hand, such advancements could represent a major increase in productivity and profitability in the world, enabling societies to guarantee food, transportation, water and energy provision and so on. However, it could also lead to the widespread displacement of jobs and alter economies in ways that disproportionately affect some sections of the population, in this case, more likely the disadvantaged communities. The lack of transparency about the sources of data, the functioning of the technologies behind smart cities services and innovations may significantly hinder accountability.

The ethical and social impact of smart cities services is a thriving and challenging field of study. Recent work such as Julia Angwin's study of racism in criminal justice algorithms[3] and Kate Crawford and Ryan Calo's study on the broader impact and consequences of artificial and disruptive technologies in societies[4] highlight the need to fully comprehend how data digital technologies underpin smart cities services.

Furthermore, given that many disruptive systems are increasingly relying on large-scale and sometimes sensitive datasets, issues associated with trust and privacy [328–330], ownership, bias, transparency [331] and fairness will only be exacerbated. Each of these areas is of highly complicated nature and they also intersect in many complicated ways. For example, an automated decision-making system could undermine privacy through the expansive use of personal information, or lock-in existing societal biases that appear in data used to train algorithms that will unlock city services. This poses important questions about the kind of smart cities unlocked by digital services. Any effective data infrastructure needs mechanisms of anonymization. Rather than using anonymization algorithms prone to failure, blockchain-based interactions – although a novel technology in the experimentation process – have the potential new mechanisms of anonymity or pseudonymity, increasing the difficulty of relating personal and sensitive data back to the individual [332]. Blockchain enables

---

[3] https://www.propublica.org/article/machine-bias-risk-assessments-in-criminal-sentencing
[4] https://www.nature.com/news/there-is-a-blind-spot-in-ai-research-1.20805

the automated transfer of value across digital networks without the need for interme-diaries. Blockchains are distributed and are based on a combination of encryption and peer-to-peer technology to update a common and immutable record at the time a transaction has occurred [333]. All nodes of this *distributed ledger* are synchronized with information about the transaction. Privacy using blockchains might be achieved using 'identity' models that remove identity-related information from the data at the point of collection/generation rather than distribution. This ensures the enforce-ment of privacy alongside the supply chain of city data. Prominent programmable blockchains available on the market include Ethereum,[5] open-source Hyperledger[6] and Openchain.[7]

Interpreting correctly the data collected from the urban environment and extract-ing insights from them is not a straightforward task. Biasing effects are a known phenomenon in information systems research [334], and it has been extensively found in decision-making process in urban environments. For instance, Kahneman and Tver-sky [335] demonstrated that though big data applications are fed with vast amounts of heterogeneous data, the bias effect on even a small proportion of the data is propa-gated throughout the final data analysis. Measuring the performance of cities through data can lead to exclusions due to biases that algorithms picked up from datasets or overfitting. This creates a tendency for such intelligent algorithms to generalize the bias in future predictions, responses, policymaking instead of providing equitable and non-biased solutions.

Understanding the changes and impacts following the design of digital services for smart cities in the context of their own individual society, and plan the appropriate strategies and responses to any threats, is vital if societies are to address and manage the risks and exploit the full potential that disruptive technologies can deliver.

6.2.3 Formulating a service strategy

The change of powers in city data offering is beginning to happen whether cities plan for them or not. It has been driven by the increasing adoption of social media, technology adoption, widespread mobile services, and increasing engagement of citizens with public services and their expectations on the degree of interactivity they want from city services (e.g. mobility application). As such, successful services in data infrastructures should provide many functionalities that transcend space (and time) and break down the barriers to information access to enhance communication and collaboration. It enables people to have access to information that will enable them to innovate, to work better, to commute more efficiently in between places, enable governments to get insights on the urban services being provided anywhere and anytime they want. To accomplish this shift in paradigm, cities must engage with their residents as owners of and participants in the creation and delivery of city data and digital services. Ensuring societal needs for city data is recognized as the

[5]https://www.ethereum.org/
[6]https://www.hyperledger.org/
[7]https://www.openchain.org/

starting point for city data service offering can be a powerful driver of data service transformation.

To increase customer value, cities must take human behaviour and needs as seriously as technology, understand which services and data are needed to solve social problems and drive innovation, identify what makes data and services more accessible to users, and what affects user's experience while interacting with services provided (e.g. usability, feeling of security and trust). Furthermore, **Services Extension** can lead to an increased value of services to the customer of data infrastructures. For instance by enabling developers and data tool providers (e.g. visualization tools, data cleansing, mashups) to link their solutions to the overall infrastructure in general leads to increased value of services to both customer and end-user. The extension of the platform services is determined by the platform openness strategy.

In the previous chapter, we highlighted that previous efforts have often neglected the end user's needs for finding, processing and consuming city data. As a result, those efforts have not been successful as one would expect. Thus, it makes necessary to lower the knowledge required for data access, that is, provide means to people to easily discover and share data, in order to achieve a large-scale dissemination. It further reinforces the importance of the feedback loop the platform must support in order to enable service providers (e.g. data/knowledge providers) to understand the value customers perceive from the product they are delivering, and how they can change their strategies so customers can be satisfied. An efficient data infrastructure in which users can find, understand, use, re-use high-quality data may positively impact the user's perceived value. Several surveys have identified personal information security and privacy as being among the most pressing concerns when it comes to city data.

In summary, cities should embrace and accelerate the use of value-added data and development of services that will put data in the hands of all citizens to unleash significant amounts of innovation. The following activities offer a logical means to use the concepts of the service domain which guides the design of services of data infrastructures. Through these activities and the use of the service design template illustrated in Figure 2.3, cities can elicit the requirements to deliver new digital services that will address the societal needs of cities in a positive manner.

STEPS IN SERVICE DESIGN

- Take societal needs as the starting point for data and digital service offering in smart cities.
- Promote engagement activities with the public to help individuals and businesses to learn which services should be provided by the data infrastructure. It should become an opportunity to stimulate the creation of new services and data releases in such a way it increases innovation and the development of applications aimed at solving city challenges;
- Ensure services and data can generate decisions that align with ethical norms and values of all of society.

- Implement different channels for data release, and different levels of data access are offered to citizens, public and private sectors according to the terms and conditions to which the data is associated with facilitate data findability, discoverability and the traceability of the provenance and ownership of data.
- Conduct interdisciplinary work that brings together experts from the humanities, social sciences and beyond, along with voices from civil society and technical insights to analyse the impact and consequences of sensitive services.
- Implement different channels for data release, and different levels of data access are offered to citizens, public and private sectors according to the terms and conditions to which the data is associated with.
- Implement mechanisms that ensure data services comply with relevant regulation, government open data policies and legal agreement with regard to proprietary data.

6.3   Technology design

The technology design of data infrastructures serves as the foundation for the widespread exploitation of data. It should be based on a citizen-centric, interoperable, open and innovative vision. This implies a significant change in the operational model of the current digital strategy adopted in cities so that stakeholders can collaborate, disperse data can be re-used and fragmented silos of digital assets and services can be brought together to deliver smarter services. Interoperability issues are limiting the realization of those necessary changes. To accomplish those changes, cities need to understand the existing technical and non-technical barriers to interoperability. On the one hand we must ensure systems and data complies with technical and semantic standards aimed to address interoperability issues. On the other hand, there are several non-technology aspects hindering the effective interoperability of systems and data, including different strategies to manage personal data, to exchange data among different stakeholders and licenses terminology.

Interoperability problem arises whenever two systems cannot interoperate because of the different data structures used in the exchanged data. Data interoperability deals with the agreements on the format and the meaning of the data that allow different systems to work together. For instance, when one legacy system is replaced by a new system with the same functionalities, it is expected that the later will use different messages than the former. Additionally, uncommon requirements of some applications might require the design of a specific hardware platform which will further increase the overall heterogeneity level. In such a case, a straightforward solution is to develop a data infrastructure to meet each device requirements and application to be deployed. Nevertheless, whenever there is a change in the requirements or a new device is integrated into the existing system it is required exhaustive programming effort and excessive resource investments in order to accommodate the changes. When such modifications are performed automatically, not requiring any programming effort, and ensuring that the overarching system will work in an orchestrated manner, the problem of data interoperability is alleviated.

The technology domain focuses on a technical architecture and its functionalities. Together they can shift city data management away from a silo-based model towards an integrated, scalable, multi-channel and engaging service delivery approach: an approach that enables cross-domain city data and enables decision-makers to take advantage of unprecedented insights into how the city and its infrastructure functions.

Basically, the technical architecture consists of applications, devices, service platforms, access networks and backbone infrastructure. Together with the service design, the technical architecture serves as a guide to the technical design [320]. Requirements as defined in the service domain determine and specify the technical architecture, which is part of the technology domain. In the technical architecture, applications, services platform, extensions, open API's and interfaces play an important role, in addition to network and infrastructure characteristics, in facilitating the process that enables the data infrastructure development, creation, delivery, bundling, control and management.

In the first chapter of this book, we have highlighted the issues associated with the centralization and fragmentation of data, and cities' wide range of disperse deployments of technology components from multiple vendors, with their own technological approaches, standards and policies. The implication is that each solution/application may interoperable on its very own and original setting, and that systems replicated from them may not be compatible and mergeable with larger strategies. Kaisler, Armour, Espinosa and Money [424] conclude that data diversifies and multiplies at unprecedented and unplanned speed, requiring ever bigger and multiple storage facilities and diverse and combined analytic techniques, while engaging different actors who tend to lack knowledge of each other let alone collaborate.

During the past decades, decisions about city infrastructures and services have been made based mostly on traditional structured data stored in relational databases. Cloud computing technologies have played a key role in enabling ubiquitous solution for managing sensory data and deploying novel business models which were not economically feasible in traditional municipalities infrastructure settings [336,337]. Cloud computing also reduces solution's time to market and transfer the risks of, for instance security and systems failure, to the provider of the service. In smart cities, cloud computing alongside machine learning methodologies have enabled the rapid proliferation in the number of applications which leverage various cloud platforms, resulting in a tremendous increase in the scale of the data collected, consumed and distributed in the urban environment [336,337].

The concepts in the technology domain and their relations are mapped in Figure 6.4. In this figure, the grey boxes within the grey container represent the concepts of the technology design and their relationships, whereas the outside boxes represent concepts from other domains (service, value network, value (societal and financial), governance) which impacts/influences the concepts of the technology design.

In this figure, the **technical architecture** is composed of the following technological components: applications, services platform, extensions, backbone infrastructure, open API's and interfaces. The stakeholders (consumers and end users) of the platform invest and/or put requirements into the technical architecture. Data

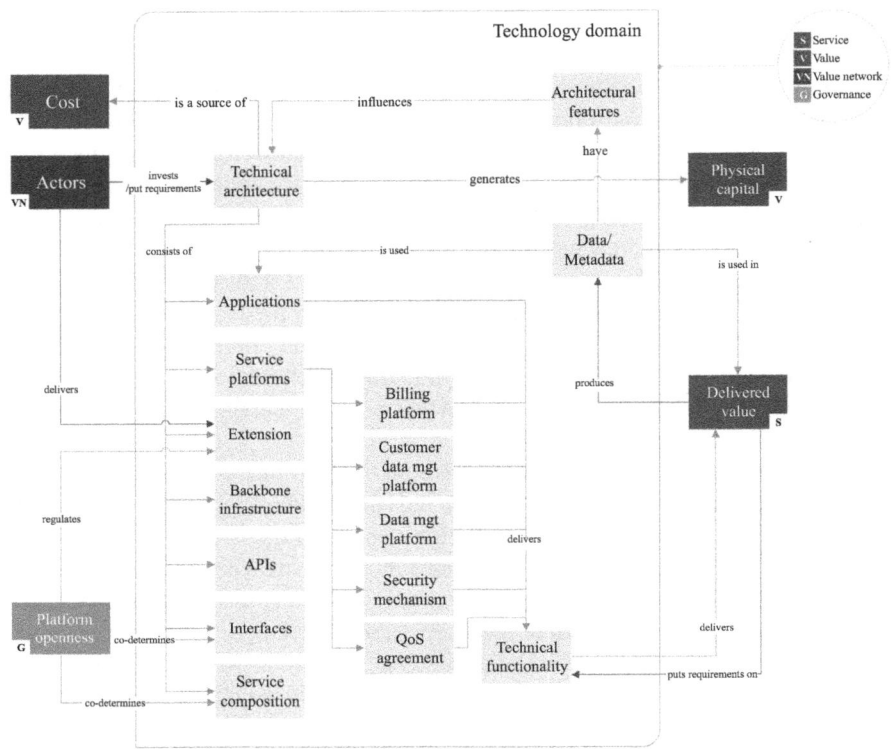

Figure 6.4   The technology design (adapted and extended from [271])

and metadata also put requirements into the technical architecture due to the different architectural features of the data. Below, we provide details for each of these components.

Applications. The applications component refer to the end-user applications providing access to city data. For instance, there is a need to incorporate data quality components directly into the architecture in order to enable data publishers to verify the consistency of the data being submitted for publication.

Service platforms. The platform services refer to the middleware software enabling different functions, including Billing, Customer Data Management, Data Management, Security Mechanisms, Machine Learning Services (e.g. Google TensorFlow, Keras, Service discovery and composition as well as Quality of Service Agreements (QoS)). For instance, user profiling can guide service providers' future strategies as they provide feedback on user behaviour and experience while using the services. In addition, if users are billed for consuming city data and/or content the user must be identifiable for service providers. Given risks of identity theft and fraudulent behaviour, security is a key issue in any authentication process. Important characteristics of service platforms are distributed platform, with personalized services, secure and open.

With regard to Security Mechanisms, whilst assisting in solving a diverse set of urban problems, embedding digital technologies in smart city's environment leaves cities vulnerable to other issues. In particular, it has the potential to create poorly designed and exposed city services and spaces that are prone to viruses and security hacks that can cause severe civic threats and issues. [338,339]. Software-enabled technologies are prone to failure, and the networked and distributed ones often receive regular updates to cope with new technologies, environments, protocols and others. As these software systems become tightly integrated it creates an even more complicated, interconnected and dependent network of software systems that is challenging to secure its stability, robustness, maintainability and security aspects. For instance, the Israeli government reported that its water, electricity, banking, transport and road infrastructure has become a target of numerous cyber attacks, and in 2013 alone the attacks amounted to 6,000 attempted hacks every second.[8]

Extensions. Services extensions are applications that provide additional services to users. For instance, it could be serviced to augment data, integrate, assess the quality of data, applications to convert data formats, among others. To accommodate extensions the platform should be easy to extend and evolve.

Backbone infrastructure. It refers to the medium and long-range backbone network infrastructure. Important characteristics are high vs. very high bandwidth, future-proof vs. non-future-proof. To define a backbone infrastructure for the platform we first need to analyse the architectural features of data and its usage.

Open API's. The availability of Open API's supports the view of fostering open innovation and crowd-sourcing in smart cities, which means that developers and interested citizens, residing outside the boundaries of the platform/organization, openly utilize the resources provided by the platform. The resources can be offered through open and well-documented APIs (e.g. standardized interfaces, contracts), which can be fully controlled.

Interfaces. Services interfaces (accessible) enable users to have access to the platform services and data facilitators to integrate their services into the platform.

The **technical functionality** of the data infrastructure is delivered by its technical architecture, whose functionalities are often presented as a functional requirements list. Some of the technical functionality characteristics are always on vs. time-critical, personalized vs. non-personalized, secure vs. non-secure. The platform must support time-critical processes, data processing services and guarantee a safe data environment. In order to provide an efficient platform to deliver cities', data infrastructure some capabilities must be guaranteed such as open standards, scalability, cross-platform support and data interoperability. Besides, the platform has to work continuously, that is, providing live-services. The technical architecture of the platform is part of the physical capital of smart cities.

---

[8] https://thehackernews.com/2013/10/israeli-road-control-system-hacked.html

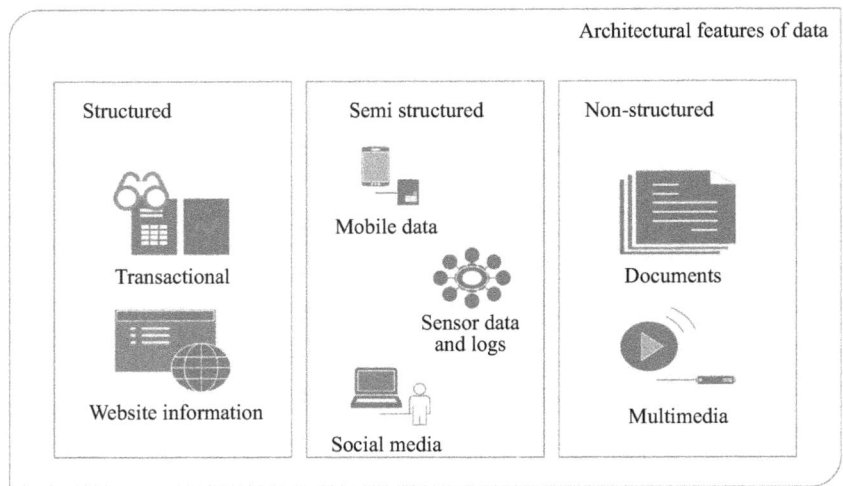

Figure 6.5   Examples of data architectural features

Another very important component of the technical domain is **data/metadata**. It is desirable that a platform for city data management provides data in open formats (public specifications) that is both human and machine-readable and understandable. As the city data is expected to not have a defined schema, metadata is of utmost importance to help machines to extract the meaning of the data. Data usability is a prime differentiator and it is a valuable competitive asset that increases efficiency, increases the service perceived value of the customer/end-users and drives profitability. The data usability can be understood as how well the data is structured and how easy the relationship between data is understandable by different applications, services and humans. City data presents high volume, different velocity requirements (collection, storage and retrieval) ranging from batch/historical data to real-time data, different formats and quality requirements.

To elicit the requirements for the technical domain of the data infrastructure, it is necessary to understand the **architectural features** of city data, the services provided and the usage patterns of the data infrastructure users. City data is composed of data with different architectural features [340,341] and they are outlined below and illustrated in Figure 6.5.

Architectural features of city data

- **Structured data**: This refers to data stored in agreement with a strict database schema such as respecting various attributes of data, delimiting the scope, domain, type, etc. This is the case, for example, of data stored in Relational Databases (RDB).

- **Semi-structured data**: This refers to data with a loosely implied schema in which it is not always possible to predict all aspects of a given piece of data. Some attributes of this data can be defined in advance and others are added at a later stage depending on the circumstances. Mobile, social media and sensor data and logs are examples of semi-structured data.
- **Non-structured data**: This refers to data for which no scheme is specified, comprising just of contents and a means of presenting it. Documents and multi-media files are examples of data with no defined schema.

Structured data are expected to be gathered from traditional relational databases used by urban systems (e.g. transport, water, energy). Citizens will be responsible for a great part of the huge volume of multi and non-structured data expected to arise in cities. This data will mostly come from web blogs, social media, mobile devices, automobiles, smart cards, smart homes (Home Area Network – HAN) and others. Multi-structured data are also expected to be generated by Governments (through the release of open data, e-government services), businesses, service providers and operators (e.g. utilities, transport) and from citizens' crowd-sourced data.

6.3.1 The 9 v's of city data

Above and beyond different data structures (which will be referred here as data variety), big data has five characteristics which must be taken into consideration and they are volume, velocity, variety, veracity and value. These characteristics are known as the 5 v's of the big data which was later adopted by Gartner [342], IBM [166] and others. These characteristics discussed in previous works are provided in the following paragraphs. We argue the 5 v's of big data are incomplete when it comes to fully describe the characteristics of city data.

Through the results of this set of analyses, we have expanded the characteristics of big data by introducing four new important data features and expanding the definition of value. The four new characteristics are variability, vulnerability, validity and visibility. The current notion of value often takes into account only the benefits brought by the use of data. We have expanded this simplistic notion of value to *value from consuming data* and *value from supplying data*. We identify and discuss the characteristics of city data based on the 9 v's of city data which is presented below.

Volume. Volume refers to the huge amount of multi-structured data being created [166]. The definition of the big data volume is subjective and it is not defined in terms of being larger than a certain number of terabytes and it also depends on the type of data, but 100s of petabyte/terabyte of multi-structured data may be named big data. According to Cisco report, "The Zettabyte Era" published in 2010, by 2016 the global Internet traffic will reach 1.3 ZB per year and IDC's

2011 Digital Universe study[9] estimates this data will grow by a factor of nine in just five years.

In the smart city scenario data volume is driven by the realization of the Internet of Things, the release of Government Open Data (GOD) and crowd-sourced data. City datasets can either focus on aggregated data or be at an individual unit level, that is, aggregated and raw data. The volume of city datasets can also vary considerably, with only a small proportion being truly classed as Big Data, although this may not always be reported in the statistics. Chapter 10 of this book provides examples of volumetric estimation of city data.

Velocity. Data velocity can be understood as how quickly the data is arriving and stored, and how quickly the information can be retrieved and processed to meet the demand. Zikopoulos [166] defines big data velocity as the speed at which the data is flowing. The constant flows of big data that must be moving at a rapid pace are not manageable by traditional systems. Some of the datasets studied in this research provide data as live feeds via APIs.

Time-sensitive processes in a city, such as fraudulent operation, traffic flows, infrastructure monitoring, emergencies or tragedies, should preferably be detected before it actually happens. Hence, data that could be useful for time critical processes should be provided as live feeds and used as it is streamed in order to maximize its value. A smart city platform will require different speeds in velocity, and therefore, robust software analytics tools will be required to intelligently find prompt and efficient ways of extracting useful information from a big pool of data to facilitate decision-making process.

Veracity. Urban raw data is often not verifiable/verified nor validated until processed for a specific use. Most of the times data analysis cannot be performed multiple times as data keep growing and/or changing constantly. The veracity of smart city data is concerned with the uncertainty surrounding the enormous volume of data coming from many different sources and with different velocities. For instance, high-frequency data collections from sensors result in a large number of data points, each of them, possibly, in error [343].

High-quality data is a prime differentiator that can increase efficiency, enhance customer service and drive profitability. The cost of poor data quality for a typical enterprise is estimated to cost 8%–12% of revenues. For instance, ABN Amro was fined $80M for not having effective data quality compliance, and Severn Trent Water was fined £26M by regulators for trying to cover up data quality issues created by its data migration project (X88 Software, 2011). The veracity of data can either empower the city or lead to poor understanding, decisions and provision of services. Ensuring and maintaining data quality is a challenge that a software integration solution will be required to address in order to provide quality assurance. Several datasets found in the existing datasets were not verified as it contains several missing information.

---

[9]https://www.emc.com/collateral/analyst-reports/idc-the-digital-universe-in-2020.pdf

Table 6.1   *Most popular data formats at*
Data.gov.uk

Format	Number of datasets
CSV	4,022
HTML	1,964
XLS	1,837
WMS	1,634
PDF	1,031
XML	456
RDF	278
JSON	196
Zip	193
ODS	173

Variety. Data variety refers to different data structures which makes it difficult to analyse with traditional database technologies for their strict schemas.[10] Part of the smart city data can be less complex for having structured and standard form such as data stored in relational databases, and other parts can be raw, semi-structured or non-structured data. The latter, which is much more complex to handle, refers to all kinds of data which has no identifiable structure and is very difficult to extract information from it. In fact, the big data variety is better described as "multi-structured data". This data can be created by sensors, smart meters, smartphones, geo-location devices, RFID logs, videos, text, messages, images and so forth.

Data variety is a valuable resource of information in which businesses (or in this case, cities) can extract knowledge and informed decisions out of it [166]. Although services in a city can be improved through the analysis of this variety of data, it is still a challenge to capture all these different types of data and easily correlate their information and draw insights based on it. Currently, city data can be downloaded in a wide variety of electronic file formats. At the time of writing, there were 20,896 datasets available at data.gov.uk – a reasonable proxy for public sector information currently available to the general public, although certainly not a comprehensive source. The ten most popular file formats used to represent the data are presented in Table 6.1, corresponding to nearly 60% of the data and including a number of which are proprietary formats such as .pdf and .xls. At the early rise of the Open Data movement – late 2012 – the Data.gov.uk portal received over 84 requests of new data due to the current format not being usable. We could not find information for other datasets due to the lack of statistical information.

---

[10]http://www.fujitsu.com/uk/Images/Linked-data-connecting-and-exploiting-big-data-(v1.0).pdf

Value. Data value, also known as 'data equity', is increasing rapidly as technological innovations take hold. Actuators, sensors, mobile phones, smart cards, smart meters, the web and social media platforms generate vast amounts of structured and unstructured data. Within all this information lie many potentially profitable insights regarding modelling city services performance, spatial aspects of the city, land usage, citizens' mobility and travel behaviours, and trends [161,344,345]. The data equity is only released when the data is verifiable, trustable, easy to consume by humans and machines. The value of city data can be divided into two categories: Value from supplying data and Value from using data. More information on value is given in Section 2.4.

- **Value from supplying data**: The value created by realising city data is not only in monetary and economic terms, but also from society and transparent governance perspectives. Actuators, sensors, mobile phones, smart cards, smart meters, open government data and social media generate vast amounts of structured and unstructured data. Within all this information lie many potentially profitable insights regarding modelling city services performance, spatial aspects of the city, land usage, citizens' mobility and travel behaviour, and trends [161,344,345]. Understanding the different values that can be created through the use of city data is essential to identify the enablers and the type of data necessary to unlock a specific value. For instance, monetary and good governance values can be unlocked through the release of aggregated data, while innovative services such as apps and new business require a more granular level data that is real/near-real time and with good quality. Competitive advantage has originated from innovative value-added services on top of data, and providing opportunities for innovation and the creation of new businesses through integrated data.

- **Value from using data**: Users are able to integrate data from a variety of sectors and distribute it across different domains, value chains and stakeholders, thereby supporting intelligent decisions in the urban environment and the creation of valuable new businesses through integrated services. City data in machine-readable format can improve planning and land use issues, service delivery and citizens' quality of life while having little impact on political accountability. The data value is released when the data is analysed to reveal these insights, allowing city planners, engineers and authorities to make better decisions. On the other hand, city data in human-readable format may empower citizens to make better decisions in their lives (e.g. crime rates, gas emissions, teachers per student in city schools) and increase their participation in public affairs, which in turn may raise the level of public trust and perceived responsiveness of government actions.

Variability. Data variability refers to the different meaning and/or context of the data created in the urban environment. Data that describes the content and context of data files, e.g. means of creation, purpose, time and date of creation, is named metadata. The use of the metadata associated with the "multi-structured smart city data" such as videos, images and text can provide wide-reaching benefits and facilitate its management. Hadoop MapReduce and other Big Data analysis tools can handle large amounts of data because their main focus lies on manipulating the metadata, rather than the data itself. Therefore, metadata is highly important to integrate, bring order and meaning to enormous volumes of multi-structured big data. Figure 6.6 illustrates a simplistic example of city data integration used in the development of an app to enable citizens to use multiple modes of transport.

The idea behind this application is that a citizen would be able to discover her/his nearest transport station, the lines serving at that station which takes her/him to the final destination. The user is able to obtain additional information about the station (lifts, toilets) and other modes of transport available for the user to continue her/his journey – especially using sustainable modes of transport such as bicycle sharing (the app would be able to show the available spaces Bicycles Dockings as well as creating routes for the users). Judy – the app creator – subscribed to her city transport live feeds to obtain all pieces of data she needed. Judy decides to use two additional datasets to enhance the user cycling experience: air quality (private data) and bicycle accident data (open data). Judy found many accidents and air quality data in several data repositories. She decided to purchase air quality data which she could obtain the data in a format as similar as the one she is already using. The accident data was in a completely different format and she had to carry out manual data processing once again. In addition, Judy has to manually integrate the data sources technically, schematically and semantically. The lack of widely accepted standards for expressing the syntax and semantics of urban data can exacerbate the frustration of knowledge creators, entrepreneurs and other users of city data.

Vulnerability. As discussed in previous sections, given enough data, intelligence and power, corporations and government can make insights in ways that was thought to be impossible in the past. However, smart city data raise new challenges in security and privacy since users implicitly expect their data to be secure and privacy preserved. There is great public fear regarding the inappropriate use of personal data, especially when data is being shared with third-parties. It is of utmost importance to address privacy issues in order to get the most of the big data promise. Some examples of sensitive data found in cities are transport historical data which contains user personal details and transport usage patterns, surveillance data (videos and images), smart metering data, payment of parking and bicycle hiring.

There are several privacy and security concerns surrounding smart metering data, and most of these concerns arise from the main functions of the meter itself [346]. An additional challenge to privacy in ubiquitous environments is introduced by context awareness. The ability of the environment to detect, reason about and act in accordance to the users' context has great potential for invasion of their privacy. The disclosure of context information might allow harmful inferences such as where users are and where they are not [347]. Another vulnerability issue is the level of access granted to data. As there are many different stakeholders within a city, this issue must be addressed in order to grant access to data for only those stakeholders to whom the data is of relevance. Furthermore, when data is aggregated, different access conditions may apply to the aggregated

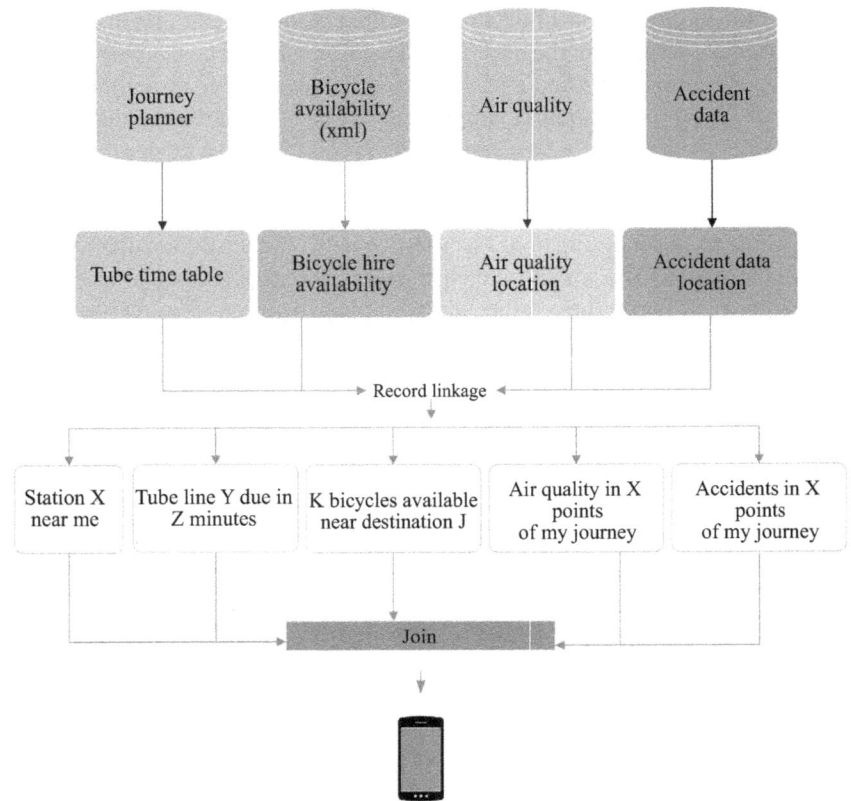

Figure 6.6   The manual data integration steps carried out by Judy

data than that found for the individual records. In addition, by centralising data in one place, it becomes a valuable target for attackers which can potentially leave huge swathes of information exposed. It could potentially undermine trust in the organization and damage its reputation. This makes it essential that big data stores are properly controlled and protected.

Validity. In addition to the previous problems, there is also the issue concerning the quality and relevance of use of data sourced from outside the city infrastructure and from its many stakeholders. This led to a much greater level of uncertainty about how trustful and accurate will be the aggregation of all the data available in the smart city (e.g. web blogs, social media, RFID logs, M2M). The extensive adoption and use of the Internet has made it easy to consume, use, shared and transform data. As a consequence, determining the origins of a piece of data is a very difficult task. Data provenance refers to the process of tracing and recording the origins of data and its movement from one place to another, enabling the examination of the "lineage" (or pedigree) of a piece of data [348].

Provenance is very significant in the smart city scenario given the open nature of the data and the accuracy and punctuality required by many city applications and core systems. The trustworthiness of data can either empower the city or lead to poor understanding, decisions and provision of services. Furthermore, platforms can use crowd-sourcing to collect information about the quality of the data which helps to enrich provenance information. Besides understanding the origin of the data, it is important to distinguish between city data that exists in a relatively static form, and city data which is continually updated.

Visibility. City data is subject to different data licensing and access policies. Stakeholders should have the right to define how users consume and re-use private data, so the ownership and copyrights of the data is guaranteed. On the other hand, open data has no restriction to its usage; however, there are also different license schemes for open data [349]. Furthermore, there is currently no linked data-based city data platform able to reason about access-control and license policies. A large majority of open data portals are publishing their data freely and under authorship (BY) or copyleft (SA + BY) models.

However, although these are the most common solutions, some governments might choose other possibilities, as, for instance, a freemium access to the data. Nevertheless, the unequal ability to understand the data greatly restrains the appropriation of data by a wide audience and encourages the emergence of "intermediaries of open data" [350,351]. In the UK, data produced by the Public sector is governed by the UK Government Licensing Framework (UKGLF), overseen by The National Archives. It provides "a policy and legal overview of the arrangements for licensing the use and re-use of public sector information both in central government and the wider public sector".[11] Making city data available under the Open Government License is seen by a large number of stakeholders as an effective means of removing barriers to exploiting the value of data. If there are instances where a generic Government License for charging for public sector information are applicable, such a License would need to be drafted in a way to avoid incentivising charging when there is no strong justification or leading it to become the default alternative to the Open Government License.

6.3.2 Formulating a technology strategy

There is a strong connection between the technologies that have to be implemented and used to deliver the intended services and the organizations and governance entities involved in supplying these. To a certain extent, the technology design is connected to the value network and governance design: the implemented and used technologies

---

[11] National Archives

to provide the services, and the arrangements between actors regulating their relationships, tasks, response organizations involved in supplying these technologies and extended services. To a certain degree, the technology design is connected to the value design: the "abilities", allocation of costs, benefits and risks and the value from using the provided services [352]. It should be clear that, depending on the choices that are made, the organizations involved in delivering the resources and capabilities, in terms of technology, finance, marketing and management, may vary. We discuss value network (organizational issues) in the following section, which is followed by the value and governance issues.

The technology design of data infrastructures requires efficient measures which address interoperability issues from both technology and non-technology perspectives. The following activities offer a logical means to use the concepts of the technology domain which guides the design of the technical architecture of data infrastructures. Through these activities and the use of the technology design template illustrated in Figure 4.4, cities can elicit the requirements to deliver a backbone infrastructure that will provide ready access and delivery of all city data that unpins the decision-making process in smart cities.

STEPS IN TECHNOLOGY DESIGN

- Identify what ICT infrastructures are needed citywide to support the data infrastructure with potential for re-use and identify gaps and strategies to fill them.
- Perform a city data mapping exercise to develop a picture of the architectural features of city data, including sources, volume, variety, temporal factors and sensitivity to identify resources needed to manage data supply.
- Identify existing standards (e.g. data, metadata, ontologies and vocabularies) and identify those with potential for re-use in the data infrastructure.
- Explore the vulnerability aspects of city data (e.g. volunteered citizen's data), and define data management strategies which ensures data integrity and compliance with National, European and International (where possible) data protection regulations.
- Put in place data licensing agreements and commercialization policies for the use and re-use of city data, following well-established regulations (e.g. UK Government Licensing Framework (UKGLF)).
- Identify integrated approaches to design and service delivery in data infrastructures which ensures that appropriate standards are adopted and put in place (e.g. Standards Hub, IEEE P2413, W3C Semantic Web Standards).
- Mobilise collective knowledge and innovation in smart cities by adopting interoperability measures which facilitates linkages with other local, national and international initiatives.
- Define a modular based architecture which is simple enough to be comprehensible at least at a high level of abstraction, identify which functionalities can be re-used by services and external applications.

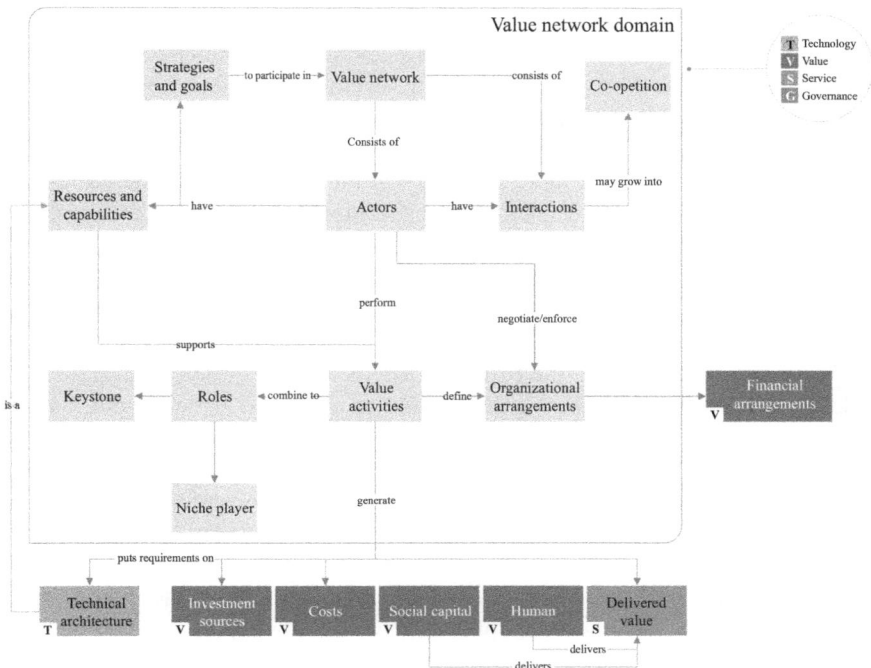

Figure 6.7   The organization design (adapted and extended from [271])

- Ensure the technical architecture relies on stable and well-defined interfaces to ensure interoperability between the platform, services and the applications provided by services complementors.
- Provide open, shared and documented interfaces to reduce entry barriers due to increased transparency and integration capability so that service complementors are able to target opportunities around the data infrastructure.
- Guarantee the data infrastructure is evolvable and able to accommodate additional functionality at the later stage at a fair and transparent cost.

6.4   Value network design

As in any SoS, the data infrastructure of smart cities has multiple stakeholders, each one with different interests, objectives and expectations in respect to the SoS. They have levels of interdependencies, social and political complexities, and competition in terms of objectives and values [353–355]. The high-level objectives/expectations or goals of each stakeholder "*() can express the rationale for proposed systems and guide decisions at various levels within the enterprise*" [351]. The concepts of the value network domain and their relations are mapped in Figure 6.7.

In this figure, the grey boxes within the grey container represent the concepts of the value network design and their relationships, whereas the outside boxes represent concepts from the other domains (service, technology, value, governance) which impacts/influences the concepts of the value network design. The main component in the value network domain is the value network, which consists of several actors and their interactions. Actors have strategies, goals, resources and capabilities. They perform value activities which together with organizational arrangements are combined into roles. The organizational arrangements affect both interactions and financial arrangements of the actors. Value activities set requirements on the technical architecture, and generate investment sources, costs and delivered value. Security aspects can be included in the strategy of an actor.

According to Selz [356], the main characteristic of the value network domain model is a value web broker acting as a central co-ordinator, an endeavour to come closer to the final consumer, and an integration of upstream activities, which is co-ordinated either with market platforms or with hierarchical mechanisms, or via a mixed form, that is, networks [357]. Markets, hierarchies, networks and information technology are integrated into an intricate web of relationships to make value webs possible [356]. Partners make up the core of the network, while contributing and support partners connected to the network are loose. As firms create products and services and engage customers in value exchanges, partners play an important role and require careful management [358].

Organizational issues revolve around the resources and capabilities, mainly related to technology, marketing and finance components which enable the service. Even if one organization provides the service, it typically needs to collaborate with other organizations to be able to provide the needed resources and capabilities.

6.4.1 Data-driven value network

The concept of data-driven value network (DDVN) is based on a platform-centric approach, which enables the pooling of multiple organizations' knowledge bases. Android and iOS are examples of organizations that have successfully become platform intermediaries and which now lead the telecommunication industry. Even powerful organizations like Google and Apple need to collaborate with the various members of their value networks (e.g. developers) in order to provide unique and inventive services and applications to end users.

A DDVN integrates processes and data in orchestrated supply chains which enable collaboration and 'co-opetition' in the provision of city data and services, as well as ensuring a response to demand that creates innovation and value. It consists of actors, interactions and the resulting relations among them. Members of the DDVN create strong relationships which collaborate for increased trust and commitment within a data infrastructure.

There are three basic types of partners in a DDVN: structural partners, contributing partners and supporting partners with varying degrees of power within the value network, based on their resources and capabilities [270]. This network of partners can potentially bring insights about specialized domains and different application markets that one single organization or city government developing a data infrastructure for smart cities would struggle to maintain in-house. Basically, this concept drives cities to deeply specialize in their core competence – governance. As a consequence, cities are able to decentralize their city data portals and platforms to create specialized supply chain networks which involve many partners, forming a large ecosystem of expert collaborators.

Architectural features of city data

- **Structural partners**: They have the most power in the network and are often formed by the platform providers and advisory partners (e.g. regulatory and standardization bodies).
- **Supporting partners**: They are the providers of service and data within the data infrastructure, such as open and proprietary data providers, application/tools developers, providers and recipients of feedback.
- **Contributing partners**: They are providers of feedback and creators or knowledge and insights, such as end users, data integrators and knowledge creators.

Government – assuming the role of provider or manager of a city's data infrastructure – must make decisions about what expertise should be provided in-house and what is left to supporting partners. This means carefully assessing the opportunities that arise to enter complementary markets (e.g. commercial exploitation of city data) and making use of mechanisms they have at their disposal to stimulate innovation within the ecosystem. This can include disclosing technical and data architecture details (as we ourselves intend to do), sharing and outsourcing expertise, creating partnerships and integrating supporting partners' solutions into the infrastructure itself. Figure 6.8 illustrates an example of a data infrastructure value network for smart cities. In the next sections we discuss the most important organization variables and their respective characteristics.

Value activities: These are the activities that an actor performs in order for the value network to deliver the proposed service. Value activities can be seen as costs but also as a source of investment. In city data services, value activities include data creation, integration, distribution, the creation of high-level streams of knowledge and innovative services. Therefore, value activities are not just the activities of the provider of the data infrastructure (e.g. local government). Furthermore, in a social-technical platform end users do not only consume data but rather, they are part of the data processing and part of the production process, value activities and the value network. Value activities, like users forming connections and interactions with actors, and creating innovative and value-added services create social and human capital of smart cities.

Resources and capabilities: These can be financial, social, organizational and technical in nature. Relations evolve over time, as trust and reputation are built.

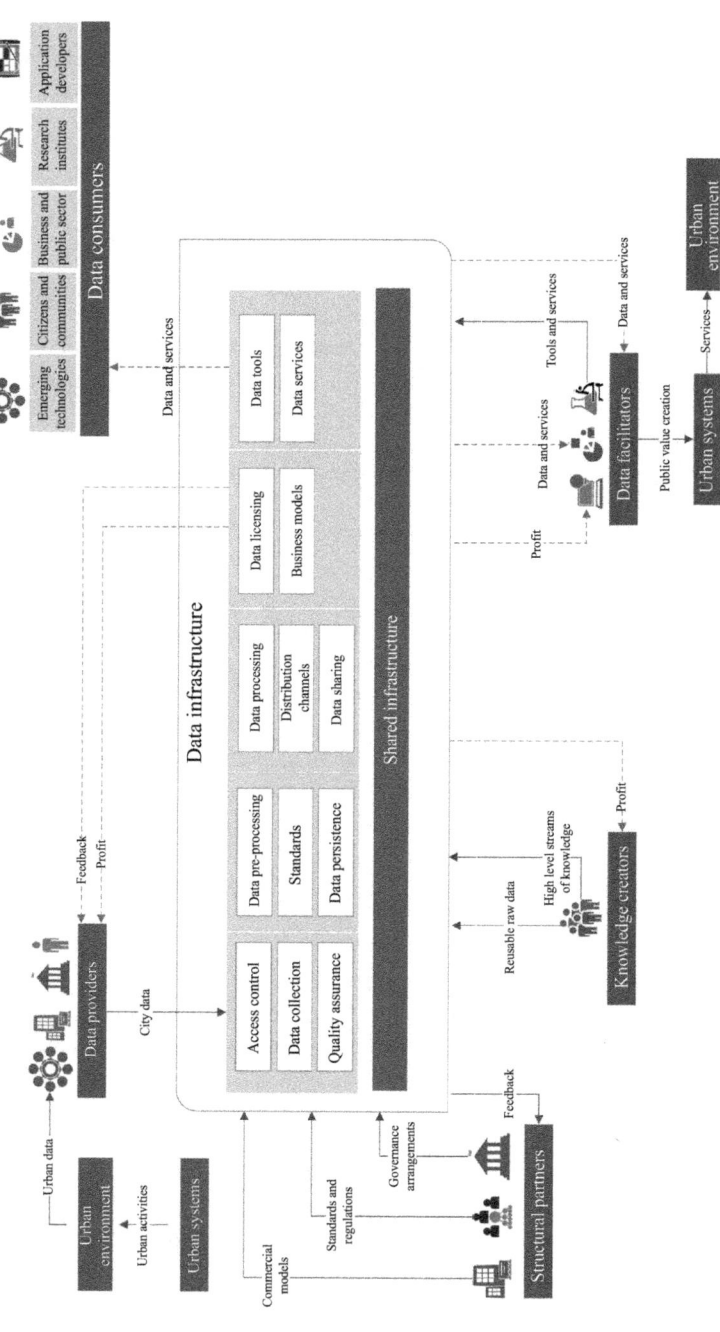

Figure 6.8 Value network design for data infrastructures

Relations are relevant not only between the organizations in a value network, but also between all stakeholders. Trust creates one form of lock-in in a community [359]. It is in the value network domain where the common data service platform needs to provide the most effort in order to accomplish the goal of providing city data services. It is necessary to look at its own role within a network of different actors and at the resources its stakeholders bring to the value network. Once this is clear, data providers and data consumers need to be identified and their specific strategic insight as to why they would want to be part/interact with the platform. It is of utmost importance to identify any value the platform adds to its customers, and in case no value is identifiable, the business case will not be viable. The common data service platform should provide ways to allow open and proprietary data providers to co-exist and cooperate in such a dynamic environment. By allowing all types of stakeholders to be part of the ecosystem, smart cities can offer more complete, specialized and efficient services to the end users.

Roles: Actors – or Stakeholders – play a crucial role in platforms as they engage in relationships and provide for interaction, exchange and innovation with and among each other. Actors can participate within one or more platforms, for example, by developing complements for a platform, or shaping and orchestrating them. Efforts have been made to classify actor groups that share similar characteristics. Following [229,230,250] classification of stakeholders, Table 6.2 presents our classification of stakeholders who interact with the data infrastructure. The platform leadership strategy adopted by smart cities policymakers may hinder the emergence of dominators. Figure 6.9 illustrates examples of stakeholder roles in data infrastructures.

Strategies and goals: These can be financial, social, organizational and technical in nature. Relations evolve over time, as trust and reputation are built. Relations are relevant not only between the organizations in a value network, but also between all stakeholders. Trust creates one form of lock-in in a community [359]. It is in the value network domain where the data infrastructure needs to provide the most effort to accomplish the goal of providing city data services. It is necessary to look at its own role within a network of different actors, and at the resources its stakeholders bring to the value network. Once this is clear, the members of the value network, their specific strategic position and their motivation to be part of it must be identified and well understood. It is of utmost importance to identify any value the data infrastructure brings to the members of its value network, and in case no value is identifiable, the business case will not be viable.

The data infrastructure should provide ways to allow all users within the platform to cooperate and co-exist in such a dynamic environment. By allowing all types of stakeholders to be part of the ecosystem, data infrastructures can offer more complete, specialized and efficient services to the end users.

Interactions. Relations may evolve from competitive to co-opetitive interactions. Relationships are important to a value network, because they contribute to trust and commitment within the network. Stakeholders' interactions with the data

Table 6.2   Stakeholders roles in data infrastructures

Format	Influence	Main activities	Examples
Keystone	High	Provides the foundation for the platforms with its standards, technology or platform. Keystones are the structural partners and create value by providing a core technology (e.g. a platform) to be used by other members of the ecosystem, they provide incentives to encourage more participants to join the ecosystem of stakeholders and innovate around the platform.	Government, policy makers
Niche players	Medium-low	Supporting and contributing partners use the technology, standard or platform provided by the keystone to create business value. Although the least influential members, they are specialized in different domains, their presence is essential to ensure diversity around business ecosystems.	Data publishers and integrators, tool and data service providers, data consumers
Dominator	Medium	Progressively eliminates other actors within the ecosystem and damages the health of business ecosystems by reducing diversity, competition, and hinders the creation of innovation	Bigger technology organization through merging and acquisitions

infrastructure can be analysed by looking in more details the relationships among actors in the urban environment as it exists before the proposed platform is introduced. This can then be compared to the new configuration which includes the proposed system. The relationships can be analysed in terms of opportunities and vulnerabilities.

Strategies and goals. Actors vary with respect to the strategy and goals they pursue with the collaboration. Collaboration requires partners to share information and provide insight into each other's ways of working, hence the importance of understanding the interactions among the stakeholders. However, strategic interest may induce partners to act against what is agreed upon, and the platform governance may provide safeguards and legal agreements to create trust between partners in order to enable open and constructive collaboration among them.

Organizational arrangements. Collaboration leads to complex interdependencies between organizations, because no single partner has formal authority over another partner. Every adjustment has to be discussed and jointly agreed [360]. To govern the collaboration, actors need to agree formally and informally on how

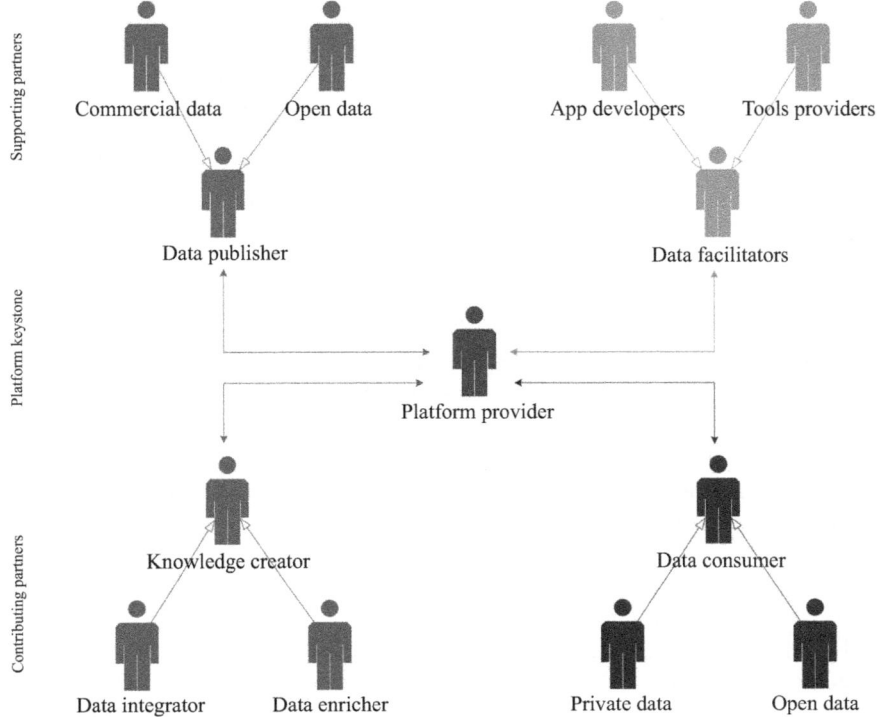

Figure 6.9 Examples of stakeholders roles in data infrastructures

to divide and co-ordinate their activities. These agreements should clearly define the responsibilities for all actors involved and should be defined in the platform governance domain.

6.4.2 *Formulating a value network strategy*

The discussion presented in the value network design highlights that the success of data infrastructures will be co-determined by the way the value network is managed and nurtured. The providers of data infrastructures must empower city data stakeholders to shape the data infrastructure and help the city to identify the requirements needed to create a prosperous city information marketplace.

To conclude from a design perspective, specifically with regard to smart cities' data infrastructure, access to critical resources is the key element in deciding which actors to incorporate in a value web. Critical resources for value webs are access to infrastructure, to content, to content developers, integrators, hosting providers, to software and application platforms, to API's, to users, customer users, billing, users support and management. Some of the resources may be found within a single organization, whereas for others more than one organization may be needed. Some resources may only be provided by one organization (Keystone), whereas for other resources several alternatives (niche players) may be available. A specific place is reserved for financial resources. Investments and the financial performance are crucially important in delivering value to customers and to the network.

The following activities offer a logical means to use the concepts of the value network domain which guides the design of the value network of data infrastructures. Through these activities and the use of the value network design template, cities can elicit the requirements to build partnerships to deliver holistic and interoperable solutions.

STEPS IN VALUE NETWORK DESIGN

- Seek opportunities to test city data consumption and business models for the data infrastructure in order to assess the feasibility of the data infrastructure and subsequent citywide and cross-domain data scale-up.
- Establish partnerships with major data providers from both public and private sectors which are committed to increasing the number of datasets provided by the data infrastructure.
- Instigate multiple-way communication among stakeholders, and actively solicit feedback to ensure future efficiency and economic growth in the city generated by the data infrastructure.
- Establish a set of principles, which the providers of city data can commit to, to ensure the validity, quality, interoperability, trustworthiness and value of the data being offered and used in the data infrastructure. For instance, that would include adopting open standards, and supporting the use of semantic web technologies.
- Disclosure technical and architecture blueprint details in order to share and out-source expertise, and partnerships, and integrate supporting partners' solutions into the infrastructure itself.
- Develop the data infrastructure in an iterative and collaborative manner, which includes DDVN members' involvement through research and digital engagement (e.g. social media, online tools which facilitate public participation in the process).
- Provide the necessary resources, information and attention to the members of the DDVN to ensure their ongoing participation on the data infrastructure development, and to have a formal managed stakeholder's engagement program in place.
- Clearly articulate to the DDVN members how they will benefit from participating and collaborating in the data infrastructure development, and how the data infrastructure will be delivered in a way that benefits all sectors.

6.5   Value design

A fair division of costs, revenues and investments is required to make the collaboration worthwhile for all the organizations in a value network. In addition, the revenue mechanism for customers and/or end-users needs to be defined, alongside costs (including transaction costs), pricing and performance indicators.

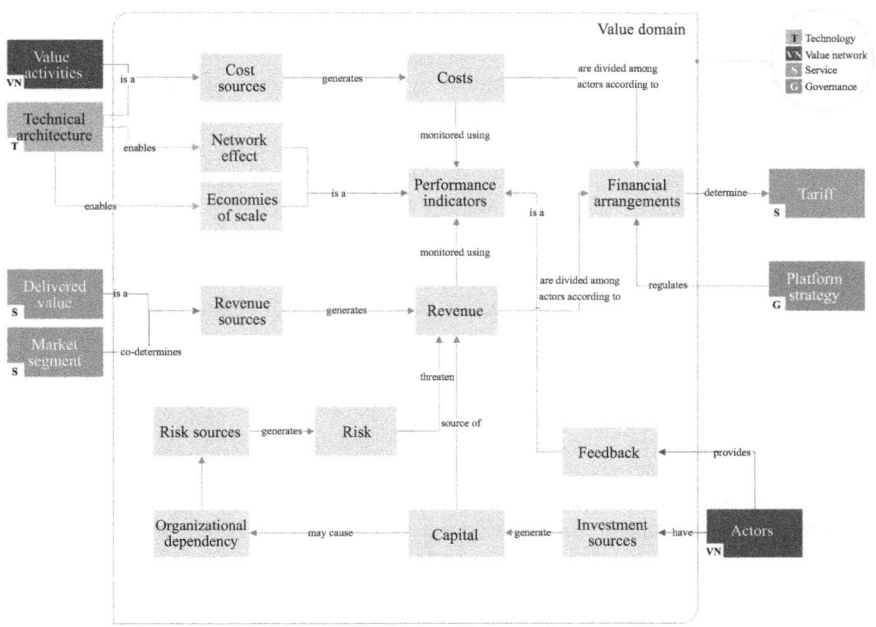

Figure 6.10   The value (societal and financial) design (adapted and extended from [271])

The concepts of the value domain and their relations are mapped in Figure 6.10. In this figure, the grey boxes within the grey container represent the concepts of the value design and their relationships, whereas the outside boxes represent concepts from other domains (service, technology, value network, governance) which impact/influence the concepts of the value design.

The **cost sources** may be influenced by the co-ordination costs of the value network. **Costs** are derived from various Cost sources, including for example technical architecture, value activities and general co-ordination of the value network. Generally speaking, the cost side is reasonably well charted. The relative importance and absolute magnitude of cost drivers will vary from industry to industry, and from firm to firm external factors [361]. Exploiting and shaping these structural factors in defining the financial arrangements is very important.

Among the various costs involved in the providing a data infrastructure, we highlight the cost of transactions. **Transaction costs** include the costs of planning, adapting, executing and monitoring task completion. It occurs when a good or service is transferred across a separable interface, for instance when a user subscribes to a data source through an API. Transaction cost economics identifies transaction efficiency as a major source of value as enhanced efficiency reduces costs. However, it should not divert policymakers from other important sources of value, such as innovation [278].

In some cases, city data is generated in the course of public and private sector activities, rather than the data being generated or collected as its core activity.

This applies, for example, to the data contained in the Office statistics on crime and Transport for London data which arises from its day to day operations. It can be assumed that such data requires relatively little investment as such data may be collected in the course of a sector's regular activities to facilitate other activities. It does not imply that its collection and dissemination is cost-free. There may be costs incurred in processing, storing and disseminating data. However, the cost of collection to support internal activities is generally thought to be lower than data requested on demand such as the data requested through the "Freedom of Information act"; however, we should also note that cost varies according to how often the data is collected. There is also cost involved in other stages of the data life-cycle: for instance, cleaning and anonymization of data.

We argue that the cost structure of data infrastructure may be comparable to cost structure of mobile services, which is characterized by a high ratio of fixed to variable costs [204] and by a high degree of cost sharing such that the same facilities, equipment and personnel are used to provide multiple services [362]. The high fixed costs typically lead to economies of scale, as increased production lowers the average production costs. Similarly, the high degree of cost-sharing leads to economies of scope, as the provisioning of a number of different services together leads to reductions in cost. Modularity in the service provisioning architecture is a way of obtaining cost advantage, as components or modules may be shared by several services. Each organization faces the question whether to perform allocated tasks "in-house" or to outsource them. Complementarity between services components may stimulate demand, whereas cost sharing helps reduce costs. For instance, instead of providing single services for data download, users could be provided with mechanisms to verify the quality of the data, and data augmentation and integration services provided by the niche players of the provided platform.

To reduces the cost of implementing many services themselves, cities can co-opting data from companies such as mobile phone operators and social media providers; however, it is not cost free: license fees may need to be purchased, additional infrastructure may need to be set up to host the data (e.g. increased cost in elastic cloud), in addition to staff to interface with the data providers in addition to process and re-use the data. An example of such partnership took place between Google and the Netherlands Organization for Applied Scientific Research.[12] This data partnership entailed using urban mobility data from Android devices to replace expensive physical road sensors for information about traffic flows. Results have shown that the mobile phone data had high accuracy and could potentially replace the road sensors. The research also reported that on a 10 km stretch of a role alone the savings would be of €50,000 Euro per year for the city. Such partnerships besides reducing the cost of implementation of physical sensors throughout cities' highway also reduced the cost of data collection as this part is outsourced and purchased from the corporation who owns the data.

---

[12]https://europe.googleblog.com/2015/11/tackling-urban-mobility-with-technology.html

The financial resources is one of the most important resources to be required by the value network. Finance also defines the bottom line of most of the services to be designed. With regard to financial arrangements, there are two main issues: investment decisions and revenue models. An important question is how investments and revenues are arranged and shared within complex value networks such as data infrastructures. Investment decisions reflect the interests of the stakeholders involved and take the mutual benefits of multiple stakeholders into account. Stakeholders that are connected through intended relationships and interdependencies consider sharing risks, solving common problems, acquiring access to complementary knowledge to be major motivators for collective investments [270]. To facilitate inter-organizational investments, organizations go through a collective decision-making process. Compared to internal processes [363] investments in which multiple actors are involved, incur high transaction costs and generate disputes. Inter-organizational investments require explicit articulation and collective agreement on the terms of investment and timing [364]. The share of each participant and the corresponding partnership ratio must be defined, that is, what each member will contribute in terms of financial and technical expertise. The success of these arrangements lies on whether or not the role of each member within the terms of institutional framework is clearly defined [364].

To can be powered by blockchain technologies that can enable *smart contracts* that can automatically transfer value across digital networks without the need for third parties [332].

Investment sources. The investments and costs of implementing and operating the service are related to the design choices made in the technology design and the question who will supply capital to cover the various costs.

Revenues and revenue sources are generated directly from the end-users consuming services and data. App developers and new businesses created through integrated services can obtain profit for the services they provide to the platform. Bouwman and Haaker [270] introduce market adoption, usage, revenue and return on investment as examples of performance indicators to help the service provider(s) evaluate and manage financial arrangements over time.

Risk sources existing in many domains may have financial consequences. The way the value network copes with the uncertainty and possible financial consequences of the risks needs to be defined. Risks in e-government services include: performance risk, privacy risk and financial risk for the consumption of invalid data, security risk involved in financial transactions, time risk and social risks.

The price and the pricing structure is typically the most visible part of the financial arrangements to the end-users in most commercial services. The owners of private data shall determine the cost of consuming their data as a service. Data enablers shall also determine the way they charge users to utilize their services. Revenues depend on the price associated with the service. In its simplest form, the price of a service is the amount of money a customer has to pay for using that service.

In an extended definition, price refers to all the sacrifices the customer has to make to obtain and use the service. In telecommunication services, switching costs can be considerable. Pricing, that is, setting prices for a product or service, is a dynamic process that takes internal and external factors into account, e.g. cost considerations and competition from alternative services.

Performance indicators. With regard to the evaluation and management of the financial arrangements over time, performance indicators, like market adoption, usage, return on investment, etc., are necessary. Equally important to financial indicators are the **network effects** and the **economies of scale**. Data and services running on data infrastructure provided by platforms are often provided and used by different groups of participants. The network effects or externalities affect the demand-side economies of scale, meaning that the demand of a service or goods defines its value [204]. Fong divides measures of success into input measures, output measures, short-term outcomes and long-term outcomes. Input measures include time and money associated with the development and operation efforts. Output measures are, for example, downloads, time spent on a website and number of transactions; short-term outcomes include adoption rates, accessibility, accuracy of information and ease of use. Long-term outcomes include cost-savings, staff-savings, trust in government and improved urban environment. Another set of measures is suggested as reach, relevance, packaging, access and collaboration, quality and operations [365].

Financial arrangements between the actors in the value network describe the way profits, investments, costs, risks and revenues are shared among the actors. Such financial arrangements are co-determined by the Platform Strategy in the governance domain.

Feedback The success of data infrastructures will be co-determined by the value perceived by users and consumers. We reinforce the importance of closing the loop in the platform by providing mechanisms to enhance the service through collecting feedback from users. The notion of *feedback* is important in open systems and refers to the situation in which activity within a system is the result of the influence of one element on another [210,211]. The implication of the notion of feedback in systems theory is that, in opening their data, governments should not simply instigate one-way communication of their data but should expect or actively solicit feedback and be able to make sense of this feedback. The opening of systems provides the opportunity for creating feedback loops in which the government can learn from the public. New governance mechanisms, capabilities and processes are necessary for dealing with these feedback loops. The nature of the response depends on the available organizational arrangements that make a response possible [210].

6.5.1   Formulating a value strategy

The following activities offer a logical means to use the concepts of the value domain which guides the design of the sources of value of data infrastructures. Through these activities and the use of the value design template illustrated in Figure 4.7, cities can elicit the requirements to create new profitable business models and develop an increased range of new and engaging services in the smart cities.

STEPS IN VALUE DESIGN

- Engage with the publishers of city data to understand purpose and cost associated with data collection so that effective ways to recover costs of opening up data can be explored. For instance, by seeking investment sources and creating alliances with the public and private sectors.
- Create new partnerships that could allow the creation of new potential and cost-effective beneficial services that could be rolled out across cities of different sizes.
- Prioritize sets of data that are in greater demand by the stakeholders, and use them for the trial of standards and new technologies, and to explore and promote use cases where data is used to deliver different forms of value.
- Support diverse business models for the open and commercial exploitation of city data (e.g. subscriptions).
- Select key partners to develop practical, citywide and fair commercial models to ensure that the way data is handled and used is completely scalable way, so as to allow data integration and exploitation to take place throughout the city.
- Demonstrate the impacts and value of city data through its independent use or re-use or by combining city data with other data sources.

6.6 Governance design

While architecture can reduce structural complexity, governance can reduce behavioural complexity. Platform owners must shape and influence its ecosystem, not direct it [207], besides respecting the autonomy of complementors while also being able to integrate their varied contributions into a harmonious whole. This is the essence of platform orchestration whose key function is to provide a context in which distributed innovation driven by value creators can emerge around a platform. It is the platform governance that determines whether innovation divisibility made possible by modular platform architectures is successfully leveraged [233,240,366]. Governance of a platform broadly refers to the mechanisms through which a platform owner exerts influence.

The introduced concepts in the governance domain and their relations are mapped in Figure 6.11. In this figure, the grey boxes within the grey container represent the concepts of the governance design and their relationships, whereas the outside boxes represent concepts from other domains (service, technology, value network, value) which impact/influence the concepts of the governance design. Below we discuss the most important governance variables and their respective characteristics.

The creation and use of powerful new disruptive technologies in smart cities requires effective governance and regulation to ensure their deployment is safe, equitable and accountable. As such, new standards or institutions may be needed

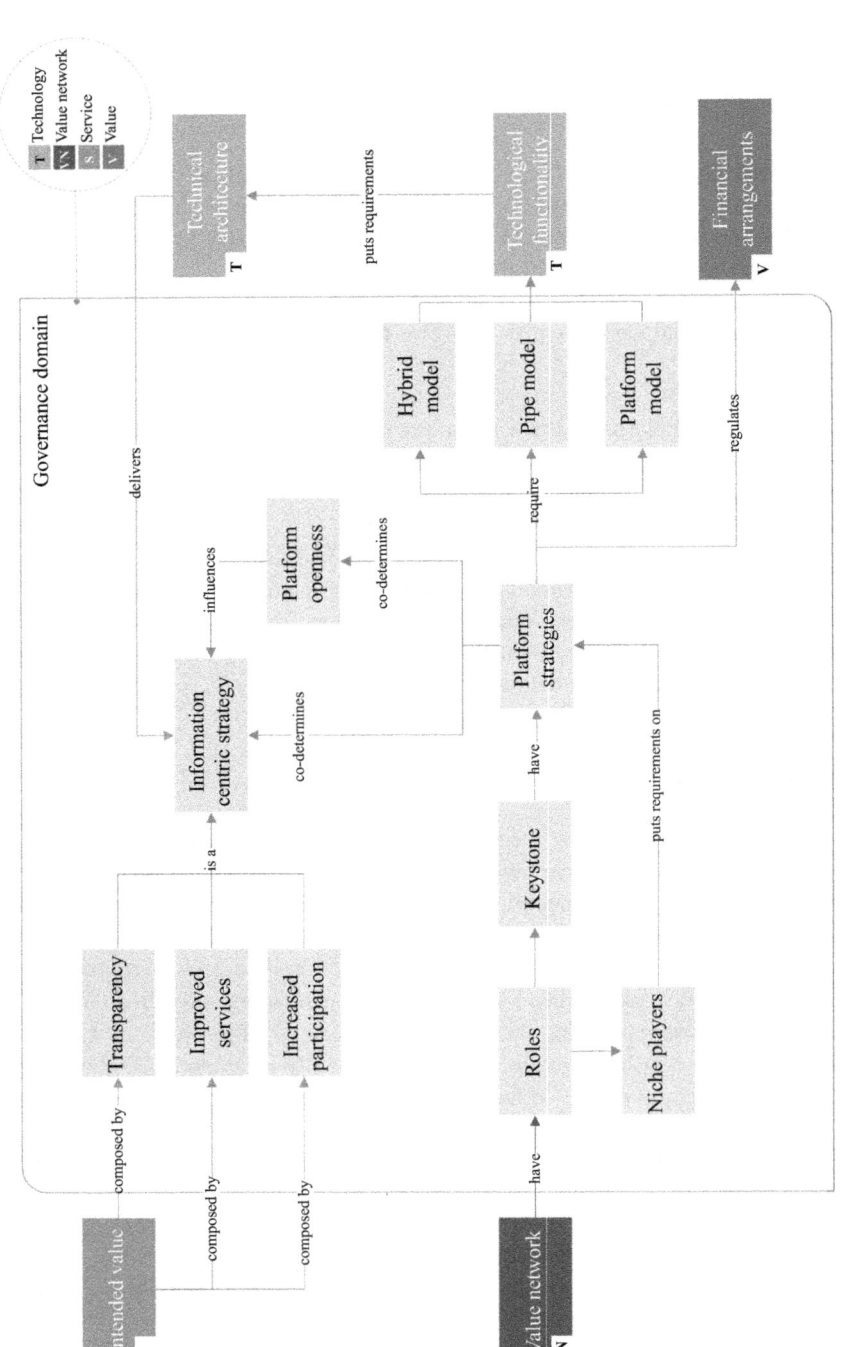

Figure 6.11   The governance design

to oversee the delivery of urban services by the public and private sector. Those parties should address important questions about how these smart cities services ought to be designed so that they are legitimate and effective, upholding the rights of everyone affected by them. Below we describe the main components of the Governance Design framework.

Information centric strategy: city data usage, especially open data, is becoming very popular and governments see this as an opportunity to increase transparency and accountability, to improve efficiency and also to contribute valuable information to the constituents. City data which is provided online, and is easily findable, readily available, accessible, understandable and usable enables new imaginative ways of interaction with the government agencies [367] including citizen empowerment [206,223,368,369]. Platform openness and governance has the power to overcome the main challenges hindering the success of data infrastructures: which are regulations and policies, data services provision, data integration and misuse.

Platform strategies: should involve platform-based competition and focuses on the development of appealing features, driving innovation by complementors, and resolving the conflicting requirements of actors who contribute to the platform by either providing data or services.

In the case of a platform for smart cities data infrastructure, we suggest the orchestration of the platform to be based on platform strategic decision rights, which means the platform owners should make decisions about what a platform should do. In order to implement such strategy, it is necessary to have a good understanding of the platform's target markets, including similar city data platforms initiatives, industry and users' needs, trends and cost structures. Therefore, it makes sense to centralize platform strategic decision rights, which also gives to the platform owner, in this case city policymakers, the power to strategically provide features which will be a differentiator from existing platforms which will attract data facilitators, knowledge creators, data providers and consumers to the platform. As in this case the platform owners retain strategic decision rights over a platform, they should retain the power to alter the rights and privileges of users and set contractual obligations and rules of participation [366]. This gives a platform owner the flexibility to make changes to the degree of **platform openness** over time.

However, centralizing decision rights with the platform owner may result in risk of overlooking a critical type of complementary knowledge that is likely to be a platform owner's weakness: deep knowledge of user needs. Such platforms are two-sided, and therefore, stakeholders will have different expectations and requirements with regard to the platform. Stakeholders of the platform must provide input in platform strategic decisions because they are likely to be able to contribute two distinct types of knowledge that are needed by the platform owner for making decisions. First, stakeholders such as data facilitators are likely to better understand their own needs for their development workaround a platform. Second, data consumers and app developers are closer to and represent the great pulse of emerging end users who will be driving the demand for city data. Whereas the Pipe model provides users with data bulk download, the **Hybrid Platform** approach model supports the provision information on demand as it shares features of a platform which provides real-time data through API's and bulk download, and therefore can fulfil the requirements of data consumers who want to consume historical data and the data facilitators who want to create real-time applications.

Platform openness: The level of openness indicates the degree to which new business actors can join the value network and are allowed to provide services to customers. The higher the desired level of control and exclusiveness is, the more likely a closed model will be adopted for the data infrastructure. On the other hand, reaching many customers may be an argument in favour of choosing an open model. Cooperation between stakeholders and institutions is very important to develop and to maintain an efficient data infrastructure for all. According to Morgan [48], active cooperation can shift the emphasis from the power to decide to the power to transform (i.e. deliver), key to overcoming the delivery deficit of efficient urban services. Gawer [219] points out that inaccessible architectures may hinder the adoption of the platform by complementors, opposed to a modular architecture with interfaces that reduces entry barriers due to increased transparency and integration capabilities [247]. Sharing technical information about interfaces will help complementors in targeting opportunities around the platform [219].

6.6.1 Formulating a governance strategy

The platform owner should retain stakeholders and make trade-offs in order to accommodate the users' requirements as well as align them to the platform governance and technical architecture. The governance domain facilitates platform-based competition and focuses on development of appealing features, driving innovation by complementors, and resolves the conflicting requirements of actors who contribute to the platform by either providing data or services. The following activities offer a logical means to use the concepts of the governance domain which guides the governance arrangements of data infrastructures. Through these activities and the use of the governance design template illustrated in Figure 4.8, cities can elicit the requirements to develop integrated approaches which ensure that services fit together and that synergies can be exploited.

STEPS IN GOVERNANCE DESIGN

- Drive changes from within city's own organizational arrangements to ensure a willingness and capability to provide city data into an infrastructure.
- Set up governance processes and usage policies for ICT infrastructure so that asset re-use the DDVN members can be maximized.
- Establish clear governance and accountability arrangements to ensure active participation, management of risks and value sources, compliance with defined rules and successful delivery of outcomes.

- Implement management strategies to continuously track the progress of the data infrastructure development, performance and feedback.
- Include public participation and consultation as an active and ongoing process throughout the development of the data infrastructure.
- Ensure the strategy is clearly articulated and it is aligned to the city's smart city vision.
- Define an optimal platform openness levels and their respective desired control, exclusiveness and target groups of services and data.
- Create partnerships with Data and Technology Ethics entities, and commit to supporting a range of public and academic dialogues about smart cities services to ensure that the services work for the benefit of all of society.

Chapter 7
The reference architecture framework

7.1 The logistical distribution of city data

In the production process, logistics can be defined as the process of strategically managing the acquisition, handling and storage of materials, parts and finished products (and related information flow), by the company through its marketing channels to maximize current and future profitability. The flow of information is vital to the success of logistics integration, since all the information (price, cost, market, etc.) must be known to all parties involved in the production process [370]. This integrated action is studied under the aegis of the supply chain.

Considering the logistics network as a large and complex framework of information, in which its success is directly related to the effectiveness in managing information flows [371], the production of city data can be considered as such, in which stakeholders combine their data and services into complex supply network providing integrated finished products – data and services – to their users, systems and physical-digital technologies.

A supply chain, in its classical form – forward supply chain (FSC), consists of all cycles involved in the manufacture of a product, involving since the production of raw materials, passing all other points of the production chain and reaching with a particular care the last point of the chain that is a satisfied consumer.

Chopra and Meindl define supply chain as

'all parties involved, directly or indirectly, in fulfilling a customer request' – and it requires a – 'constant flow of information, product and funds between different stages'

Real-world interactions in cities often occur in network structures rather than in a bilateral manner. Along this line, it can be stated that a smart city, composed of services and stakeholders, will probably function as a supply chain with strong characteristics of a supply network.

Smart city data supply chain

'A smart city data supply chain is composed of all stakeholders linked by the provision of reusable information and integrated services.'

The supply chain execution for smart cities is managing and co-ordinating the networked movement of services, information and funds across the whole supply chain. It includes, for instance, planning, scheduling and production control, service distribution planning, service delivery and so forth. When creating such supply chain, it is crucial to provide visibility and transparency of all processes and data for monitoring as the final product can get easily quite complex, that is, be a composition of other services and include many stakeholders. In the context of smart cities, we propose the following variant:

Types of city data

Open data: Non-privacy-restricted and non-confidential data produced with either public or private resource and is made available without any restrictions on its usage or distribution [192].

Proprietary data: Data that is subjected to any sort of restriction, usage authorization and licencing and may contain privacy restrictions and distribution boundaries. It is produced with either public or private resource.

Commercial data: Restricted and/or licenced data including permission, charging, privacy, publication and distribution and is produced with either public or private resource.

Sensory data: Open and/or restricted data collected by different sensors, actuators and devices owned by public and private sector, and citizens. Sensory data is often diverse in nature, architectural features, mostly location and time dependent, and different levels of quality.

City data is produced by a multitude of systems, devices and applications, and whose logistical distribution varies according to its suppliers (citizens, public and private data providers), their sectors, its distribution channels and the policies and regulations to which it is subjected to.

Sources of city data

Public sector: Data produced, collected or funded by the public sector, subject to restriction relating to location, national security, commercial sensitivity and privacy.

Private sector: Data produced, collected or funded by the private sector, which can be open or private data. In the case of proprietary data, the restrictions of usage and distribution are decided by the businesses who own the data. Examples of applicability of private sector data in smart cities include Google search data which has been capable of characterizing both size and duration of epidemics of flu [374], and Wikipedia data for dengue fever epidemics [372].

Crowd-sourcing: Crowd-sourced data corresponds to data collected and distributed by humans through the use of digital technologies and social media.

Examples of the use of citizens' data in smart cities and policymaking are the Oxford Flood Network,[1] Un Global Pulse[2] and FixMyStreet.[3] These examples demonstrate that crowd-sourced data offers access to types of data which cities without citizens' participation could never hope to provide.

The logistical distribution of city data is illustrated in Figure 7.1. In this figure, data produced within the private sector data is often from Telecom companies, the IoT, utilities companies, among others. Data originated from the public sector are commonly produced by the government, local authorities and statistical offices, among others. Crowd-sourced data are produced by humans in the cities (e.g. citizens and tourists) who share their personal information, location and utilization of services, among others, in social media platforms, websites, online forums and so on. These sources of data are commonly distributed through five different types of channels: a website, non-online channels, media and reports, rest application programming interfaces (APIs) or a secure environment for data analysis (e.g. internal systems). Data availability refers to the restrictions (if any) attached with the data.

The cost of the supply chain incurs from flowing information throughout the chain, as well as from the management and restructuring of data, information, services and funds in order to maximize the overall value generated [373]. According to Fisher [374], a supply chain performs two distinct types of functions:

- a physical function encompassing product manufacturing to the movement of goods in the supply chain;
- market mediation ensures the finished good matches customer needs.

The physical function of the city data supply chain is very important for the provision of efficient city services while the market mediation function emerges from the integrated design of these services. By extending the simple reasoning that the physical flow inputs and outputs traverse the supply chain for a broader view of the value chain, we realize that it is necessary to take into account the relationships that make this chain more competitive with lower costs, and lower risks and fulfilment of the requirements of each of its members.

On the one hand, the **producers of city data** have different requirements and expectations with regard to the collection, integration, enrichment, distribution and re-use of city data. The current production of city data in forward logistics (FL) supply chains leads to deeper implications when we consider what happens when the data reaches the last stage of the chain. Does data get recycled, re-used and upgraded, or the value created from integrated data is simply discarded and not re-used? Recycling of

---

[1] https://flood.network/
[2] https://www.unglobalpulse.org/
[3] https://www.fixmystreet.com/

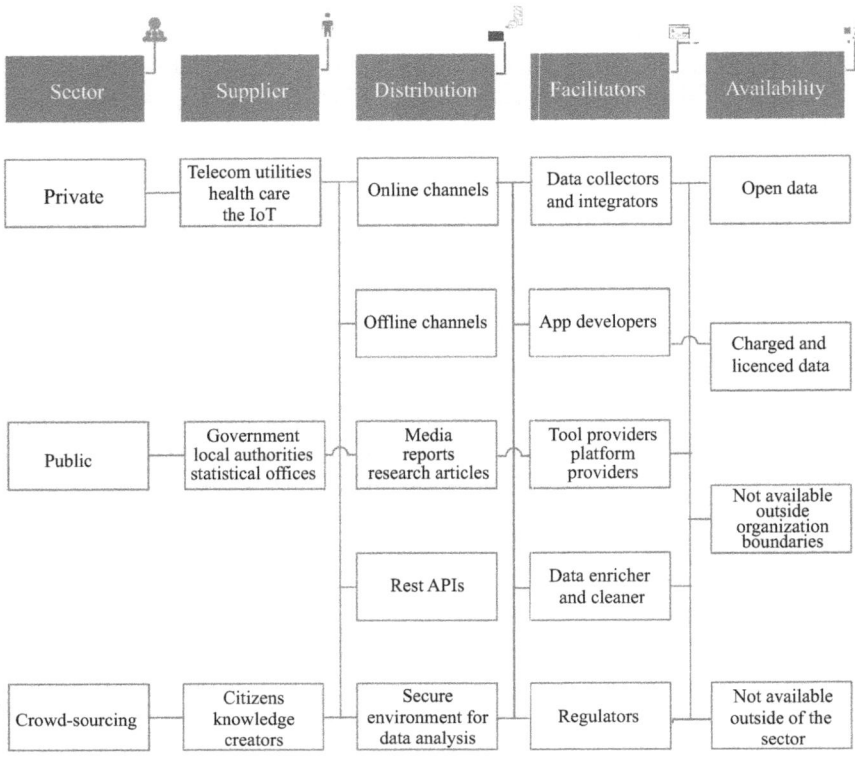

Figure 7.1   Logistical distribution of city data

data may equate to maintaining, refreshing and updating data, finding new purposes and applications for it.

On the other hand, the **consumers of city data** have expectations with regard to both consumption and re-use of the city data. City data, just like a raw resource, can flow through multiple supply chains and end up in a variety of products, making it also vulnerable to breaches. Issues in data trustworthiness, bias, ownership, security, data provenance and legal aspects of data consumption can significantly limit value creation. The FL model does not offer a mechanism to enhance data quality through collecting feedback from users. The notion of *feedback* is important in open systems and refers to the situation in which activity within a system is the result of the influence of one element on another [170,211]. The implication of the notion of feedback in systems theory is that, in opening their data, smart cities should not simply instigate one-way communication of their data but should expect or actively solicit feedback and be able to make sense of this feedback. The opening of systems provides the opportunity for creating feedback loops in which cities can learn from the public. New governance mechanisms, capabilities and processes are necessary for dealing with these feedback loops. The nature of the response depends on the available organizational arrangements that make a response possible [170].

The heterogeneity found in the FSC of city data shows that city data is not co-ordinated and is locked in fragmented organizational silos. As a consequence, the city data stakeholders are not able to deliver value into the city data infrastructure. It highlights the need for a suitable value chain framework capable of identifying the strategies to manage and orchestrate the entire supply chain of data. The flow of information is vital to the successful delivery of smart cities solutions, since all the information (price, cost, market, etc.) must be known to all parties involved in the production process in the same fashion as traditional supply chains.

While governments are increasingly aware of the impact of effective city data offering in smart cities, these impacts can be more restrained and variable than many policymakers realize. Data infrastructures depend on the context of the smart city, that is, they depend, for instance, on location, culture, available data, smart city vision, local regulations and policies. The same data infrastructure will have very different capabilities or impacts in different locations. A one-size-fits-all approach to data platforms transformation into data infrastructures and simplistic approaches to engage stakeholders of city data with one another are unlikely to work.

This network of partners can bring insights about specialized domains and different application markets that one single city government or organization developing a data infrastructure for smart cities would struggle to maintain in house. Furthermore, every activity in this supply chain affects the delivery of a final product to users, creating either positive or negative impacts. For instance, poorly managed personal data can be leaked outside government systems, unidentified biases in datasets can drive cities to focus their developments in certain areas of the city which can result in the negligence of certain portions of the population. As such, cities have the option to specialize in their core competence – governance – and decentralize their platforms and services into a complex supply network providing integrated finished products to users, systems and machines.

7.2 The forward logistics of city data

Porter's [321] value chain analysis provides a means for examining internal processes of supply chains and identifying which activities are best provided by others. The FL of data infrastructures can be described in terms of five sectors: inbound logistics, production, outbound logistics, marketing and sales and services to stakeholders. The description of each sector of the FL are outlined below and illustrated in Figure 7.2.

Inbound logistics: It involves all the activities involved in the creation and gathering of data. Urban activities trigger data creation in various systems, sensors and actuators in an urban environment. As this data is created throughout the city supply chain, a connectivity point becomes necessary to enable data collection. In the connectivity point, data can be gathered using standard protocols and

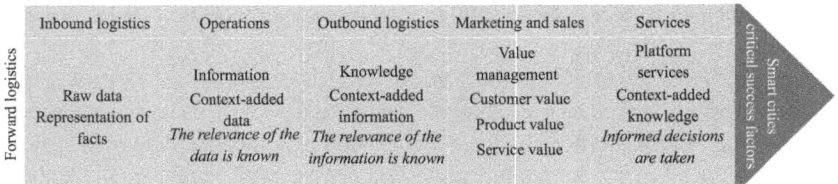

Figure 7.2   Forward logistics of city data

interfaces. The inbound logistics will deal especially with big data volume and velocity and visibility.

Operations: It involves all the activities involved in enriching the raw data gathered in the inbound logistics. Data must be pre-processed in order to verify the completeness, provenance and quality of the data. One of the most difficult tasks involves how to automatically generate the right metadata to describe the data and how it is recorded and measured. Sensitive information can be protected to assure data privacy. A difficult task in the operation silo is deciding how to organize the data for storage in such a way the data is easy to store, process and retrieve. The operations logistics will provide solutions to address data variety, veracity, variability and vulnerability.

Outbound logistics: It concerns both city- and global-wide delivery of data with added value. It involves integrating and analysing data from both internal and external data repositories. In a smart city, it is expected that multi-stakeholders will be involved in data discovery, processing and sharing. The raw data that will pass through this point of the value chain will be transformed into useful information and along with its context will be transformed into knowledge which can be consumed by humans and machines to produce informed decisions. This stage of the value chain will have to handle all the 9 v's of big data.

Marketing and sales: as discussed in the previous section, marketing mediation function emerges from the integrated design of the smart city services. In this context, it is of utmost importance to provide one-stop 'shop' for information where people can obtain the desired data. This silo of the value chain is responsible for managing the value – customer, product and service value – including financial operations (service and data subscriptions) and customer profile management.

Services: It concerns all the activities involved in delivering data and services as well as the activities that monitor if the objective of the value chain is being achieved. It encompasses open APIs, data channels, data publishing and discovery end-points, tools for processing and enhancing data and any necessary technical updates and performance evaluation.

Smart cities critical success factors: It represents the CSFs of the business models which are delivered by the value chain. These CSFs are discussed in Chapter 4 (Section 4.5.2) and they provide support for the seven capabilities of smart cities.

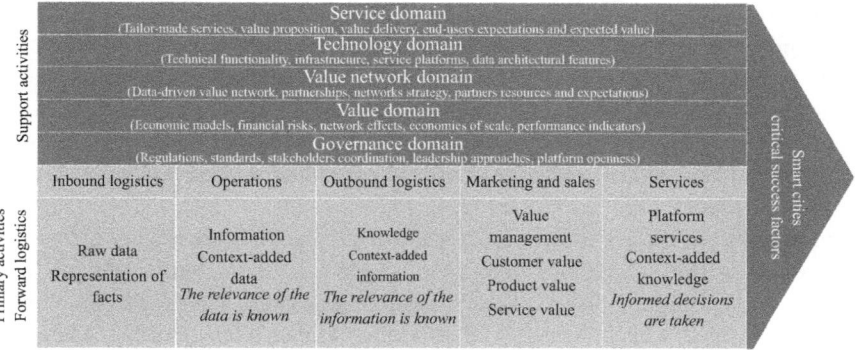

Figure 7.3   Primary and support activities in the city data value chain

7.3   Support activities

Porter [321] suggests that the activities of a business can be grouped under two headings:

- **Primary activities**: Activities directly involved with the creation and delivery of the product or service, for instance, the FL activities.
- **Support activities**: Activities (in our case 'requirements') that feed both into primary activities and into each other.

In Porter's value chain, support activities (e.g. human resource management, technology development) are not directly involved in production but have the potential to increase its effectiveness and efficiency. It is rare for an organization to undertake all primary and support activities themselves. Value chain analysis is thus a means for examining internal processes and identifying which activities are best provided by others. Figure 7.3 illustrates the combination of primary and support activities in our data value chain framework.

In Figure 7.3, the support activities consist of:

Service domain: It refers to activities concerned with the service domain of the platform, such activities include expanding or altering the scope or the services provided by the platform, as well as the quality of services related to those activities.

Technology domain: It refers to activities concerned with managing information and developing knowledge in the city. City physical infrastructure and ecosystem is the environment where the transformation from raw data to informed decisions will occur. The technological development will be derived from collaborative service design across locations and among multiple value-system participants (the city SoS).

Value network domain: It is concerned with the activities concerned with recruiting, motivating, developing the workforce of the data infrastructure such as getting new stakeholders involved, accommodating new requirements and expectations of new and existent stakeholders. It also involves cultural change in the sense of using and exploiting the wisdom of the crowds: recruiting citizens as human sensors and make use of user-generated content [375]. Additionally, it involves the technology suppliers who will provide the tools and technological infrastructure necessary to create the platform (e.g. cloud service providers, network infrastructure).

Value domain: It involves the activities necessary to maintain the finance activities of the data infrastructure, including sourcing and negotiating with suppliers. In the smart city scenario, the acquisition of data requires agreements and communication among all stakeholders, service providers and enablers. It also encompasses activities that monitor and maintain security and trust on data subscriptions.

Governance domain: It refers to platform orchestration in which the key function is to provide a context in which distributed innovation driven by value creators can emerge around a platform. It is the platform governance that determines whether innovation divisibility made possible by modular platform architectures is successfully leveraged.

7.4   Reverse logistics of city data

The evolution of the structure of smart cities towards the extended enterprise model has put some additional pressure on logistics operations. Now, it is possible to choose from among many distributions and sales channels, many points of data storage and many providers of data and services. In Chapter 2, we have discussed the importance of feedback loops from users and the need to orchestrate data, stakeholders and technology.

In FSC, new product development is defined as the creation of a new commercial product following the identification of a market opportunity relying on a set of technological assumptions [223]. Recently, attention has increasingly been paid to a set of activities taking place after the product has reached the end of its useful life, consolidated under the label of reverse supply chain (RSC) and presenting an interesting potential for cost savings related to reduced resource consumption and/or product re-use [205].

The RSC plays a central role in improving economic benefits, both financially and in terms of machine and human effort. RSC management entails the effective and efficient management of a series of activities to return products, parts or materials from the customer with the aim to recover their value [376,377]. These activities include reverse logistics (RL) and recovery options such as re-manufacturing, refurbishing or recycling [378,379].

An RSC significantly differs from an FSC with regard to its operations, management and stakeholders [380,381]. For example, while products in the FSC are produced according to market demand forecasts, the RSC is more reactive to market returns of uncertain quantity and quality [378]. Similarly, smart city data

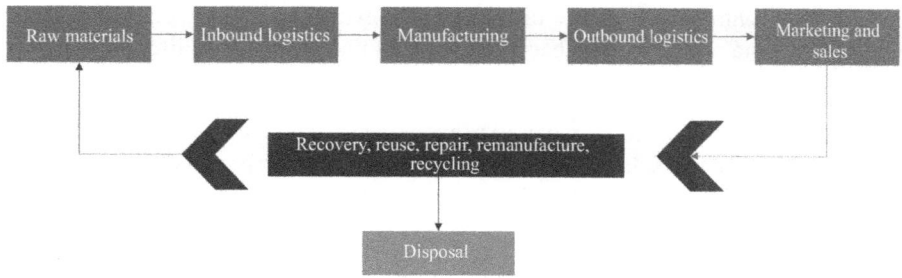

Figure 7.4   Generic FL and RL

infrastructures should, for instance, promptly react to any quality issues in data and services provision (e.g. bias, outdated information), privacy, regulatory violations.

The integration of the classic FSC and the RSC constitutes a **closed-loop supply chain** (CLSC), and they mutually influence one another because choices made in one chain influence processes in the other. In an integrated CLSC, the FSC and RSC are managed in a co-ordinated way towards the common goal of maximizing value [382]. Guide *et al.* [383] define CLSC management as

'[the] design, control and operations (of a system) to maximize value creation over the entire life cycle of a product with dynamic recovery of value from different types of return over time.'

As such, a holistic view of value creation needs to focus on the CLSC that incorporates the forward and the reverse chain, as well as the interaction between the two [384]. Managing a CLSC entails taking back and recovering used products, parts or materials for re-use in the original or a secondary FSC [379,383]. Figure 7.4 illustrates a generic form of FL/RL. During the recovery stage, possible decisions on return products are made, for example, whether to repair, re-use and re-manufacture or dispose.

The RL of the data infrastructure can be described in terms of four sectors: recovery, re-use, re-manufacture, raw materials and disposal. Most importantly, the FL connects with the RL at many points where recycled data, components/metadata, services and users feedback are reintroduced into the downstream production and distribution systems (see [385,386] for the application of RL in production chains). Consequently, CLSC may differ along crucial dimensions such as data acquisition strategies, data re-use and re-manufacturing, quality feedback, distribution and selling but also according to the life cycle stage.

According to this overall condition assessment, data/metadata is routed towards the most appropriate revalorization activities before being reintroduced into the FSC at the appropriate point of entry. The revalorized items are then re-used either by the data re-manufacturer itself or by any data provider that contributes to other data value chains. The activities in the RL contribute to minimize the negative impacts on the data environment, to enhance or facilitate the extraction of data *residual value* at end of life. Data return and feedback may occur at four general data life cycle stages:

Recovery: It involves all the activities involved in collecting feedback data from users, data that has to be repaired/re-manufactured and any knowledge derived from data existent in the ecosystem. At this stage, data can be recycled, dismantled and tested before being infused back into the platform. As discussed before, the special material flow logistics in CLSC determines the special dynamic features of product quality, which can be used to further improve the product chain of the platform.

Re-use: It involves all the activities involved re-publishing and re-use of information that is selected and ready for re-manufacturing or to be on the market. Data that are transformed, enriched and that will be stored in the platform will go to the next silo of the chain (re-manufacture), while data that is ready to re-use and consumption can be directed to the data operations and outbound phases where data is ready for consumption.

Re-manufacture: It concerns about the activities involved in repairing, disassembly and re-publishing data. In this stage of the chain, data that was repaired, updated or submitted for re-cycling are stored, enriched and linked to other pieces of data on the Internet and will be ready for consumption in the next stage of the chain that is the data outbound.

Raw material and disposal: raw data that is extracted and obtained from re-manufactured data can go back to the first stage of the chain where raw data is first gathered and submitted to the data operations that takes care of data enrichment, protection and linkage. Raw data that is not usable or is removed from the platform by its publisher is characterized as data disposal.

Figure 7.5 illustrates the integration of RL with the FL of city data. During the recovery stage, possible actions are performed over the data, services or algorithms. For instance, the platform keystone alongside supporting partners could decide whether to repair or remove a dataset or service based on complaints of service bias or unfairness or leakage of personal/sensitive information. Examples of FL activities to manage city data are acquisition (inbound logistics), handling and storage (operations), distribution (outbound logistics), subscriptions and customer management (marketing and sales) and provision of specialized services (services). In the RL activities are the collection of feedback (recovery), pre-processing and augmenting (re-use), maintenance (re-manufacture) and disposal or raw data extraction (raw material).

7.5   Inside-out and outside-in linkages

As discussed in Chapter 3, each phase of the business models is influenced by external forces which will ultimately impact the offering of city data. Hence, the business models of data infrastructures are dynamic rather than static, and it is important to

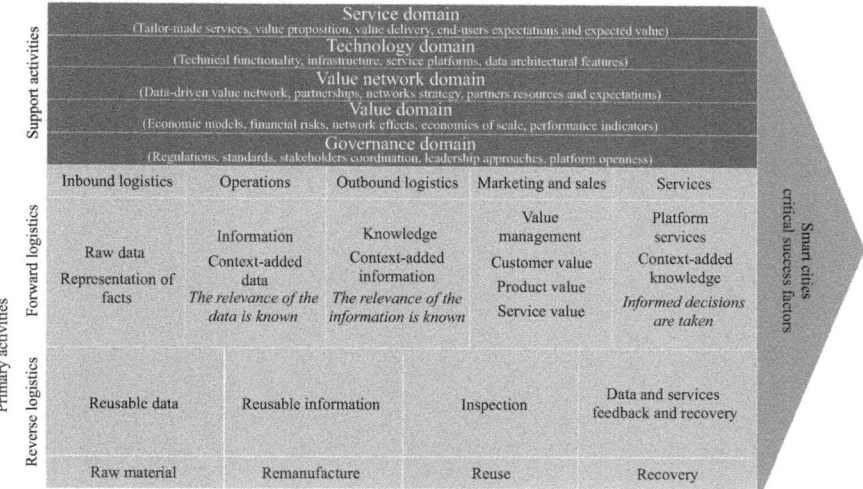

Figure 7.5 Closed-loop value chain of city data

understand how they transform over time. These external forces can be referred as to outside-in linkages [387]. Figure 7.6 illustrates the external forces affecting the four business models phases defined in our framework. Such external forces affect one's data infrastructure ability to improve operation and execute strategy.

Porter and Kramer [387] demonstrate that opportunities for big impacts of corporate social responsibility (CSR) strategy involves interlinking both inside-out – one's firm CSR level perspective) – and outside-in linkages – environmental factors affecting the CSR. Borrowing the same rationale to the context of data infrastructures, their strategic design will depend on the interdependence of the inside-out linkages – the influence that the city data value chain exerts on the smart city vision – with the outside-in linkages – the extent to which external forces affect the city data value chain. Porter [321] suggests that the activities of a business can be grouped under two headings: primary activities, which are those directly involved with the creation and delivery of the product or service, and support activities, which feed both into primary activities and into each other. Although he considers support activities as not directly involved in the production process, they have the potential to increase its effectiveness and efficiency.

Our comprehensive value chain framework illustrated in Figure 7.7 helps cities to align the capabilities of their data infrastructures with their smart cities vision and demonstrate how it helps cities to achieve maximum critical success factors, such as the partial lists of examples illustrated here demonstrates. The lack of a time dimension of Porter's value chain model could be particularly significant in the world of fast-moving technology development. Our value chain framework addresses this issue by incorporating into the model supporting activities originated from a dynamic business models framework. The value chain illustrated in this figure depicts all the

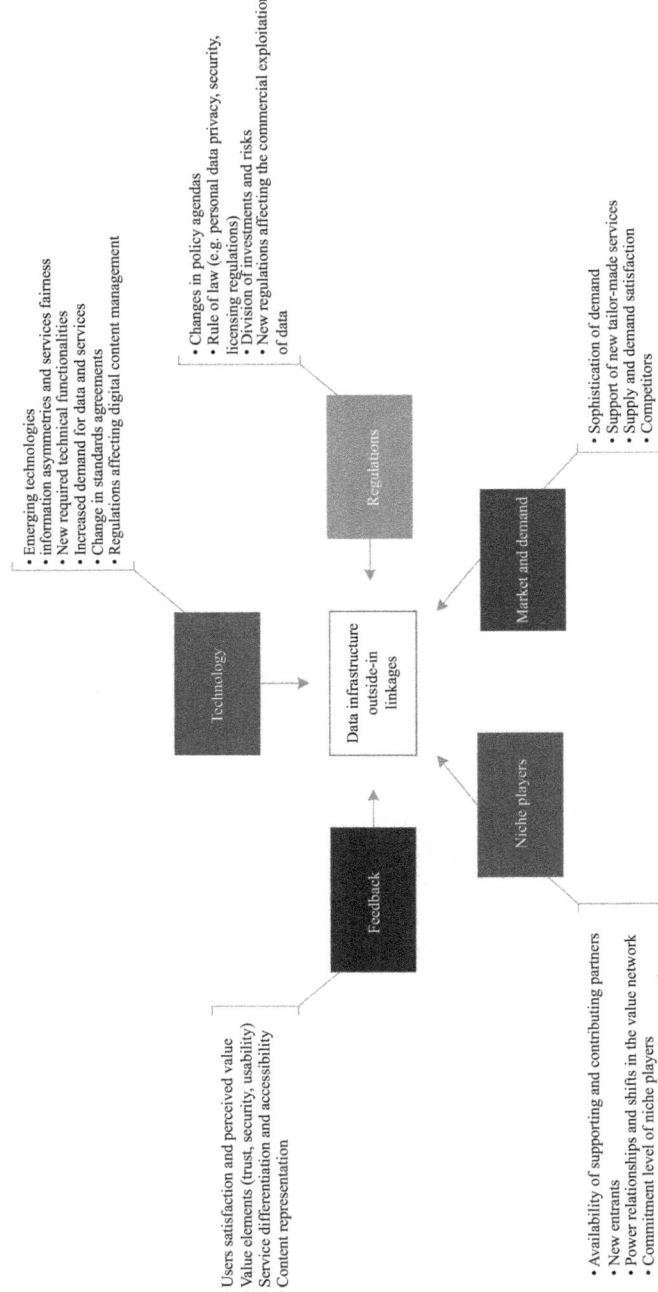

Figure 7.6   Data infrastructures external factors: outside-in linkages

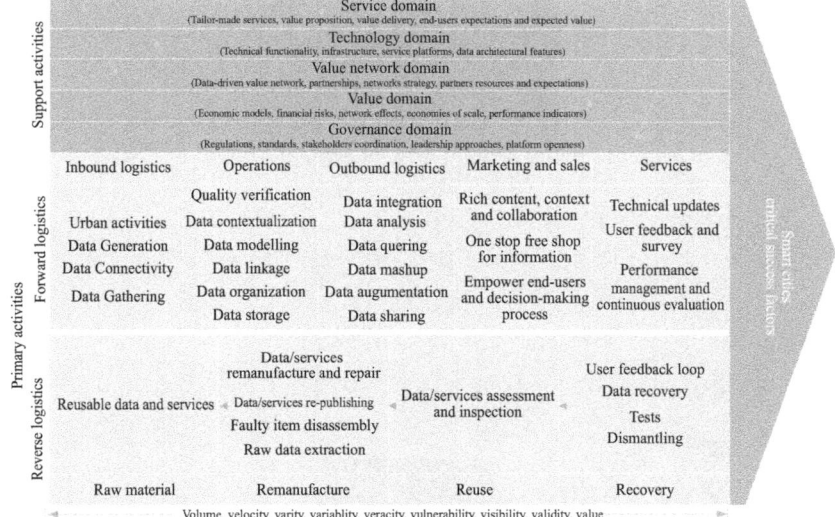

Figure 7.7   *Closed-loop value chain of city data*

activities necessary for a data infrastructure to manage city data. When cities use the value chain framework, they create an inventory of opportunities and operational issues that need to be investigated, assessed, prioritized and addressed.

7.6   Data infrastructure reference architecture

The data value chain proposed in this thesis consists of all distinct data activities inside the chain that starts on the gathering of raw data, passing through data operations to context enrich data and transform data into information, finishing on the integrated and rich content knowledge to be delivered to the stakeholders. The closed-loop framework can be represented as an initial architecture view of a data infrastructure as illustrated in Figure 7.8. This architecture view shows three important layers of the architecture: a first layer containing components specifically designed to deal with inbound logistics activities, a second layer containing components to deal with data operations activities and a third top layer containing components to deal with data outbound activities. Between data manager and the platform as well as between the platform and data, consumers reside graphical user interfaces/APIs which enable data publishing and consuming.

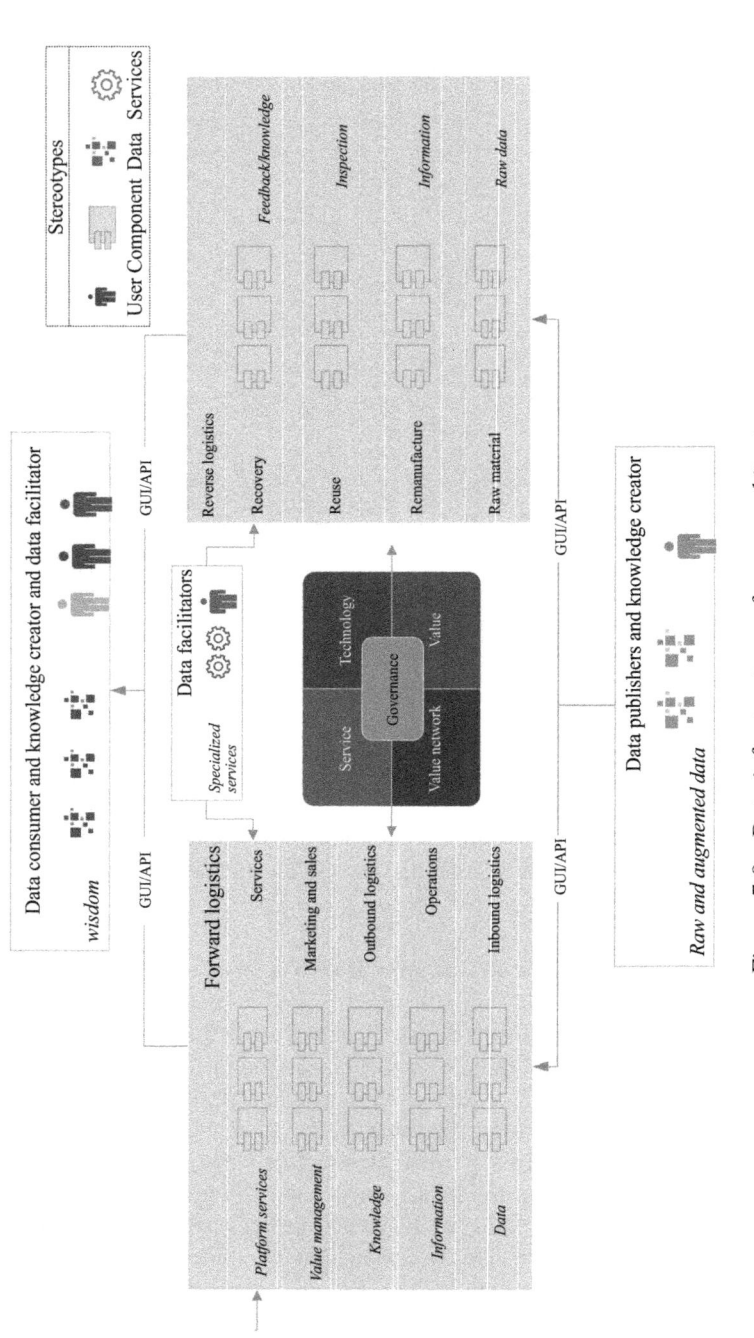

Figure 7.8   Data infrastructure reference architecture

This initial architecture view identifies the processes involved in gathering raw data and delivering knowledge that will deliver informed decisions on the urban environment. These processes were identified through the development of the closed-loop value chain. For each identified process inside the platform, there are specific requirements that will be under the responsibility of a particular platform component.

7.7 Summary

The requirements and goals of the platform for smart cities highlight that the supply chain must enable a data environment in which all stakeholders involved are able to co-exist and compete among them. Therefore, the CLSC model is designed and managed to explicitly consider activities along both the forward and the reverse flow chain [383]. Data can be produced, re-used, re-manufactured or even disposed.

The interesting point here is to facilitate the 'knowledge' created in smart cities to be infused back onto the system, and users are able to be part of the data processing and improve the platform through collaborative networking. In previous chapters, the importance of a feedback loop in which users can feedback on the quality and trust of data has been discussed, and data providers of the data/metadata may verify and provide the proper data maintenance required by users. The RL modelling in such a platform plays a central role in improving economic benefits, both financially and in terms of machine and human effort.

The value chain analysis explicitly recognizes the interdependencies and profits cost efficiencies from the exploitation of linkages between value activities of a city. For instance, changes in the standard formats (one value activity) may significantly influence the activities involved in operations and outbound logistics (another value activity). These activities must be well co-ordinated if the change in standards is to be accomplished. Establishing new data standards without the platform having means to support it will affect the reliability of the platform. Another issue to consider is a service (an activity value) not be real-time available-to-promise/capable-to-promise and fulfilment or users' needs. Time-sensitive processes in a city, such as fraudulent operation, traffic flows, infrastructure monitoring, emergencies or tragedies, need high data/service availability in order to detect any problem before it actually happens. Value chain analysis and exploration provides a powerful tool for strategic thinking for creating a smart city with a sustainable platform.

Part III

Applied data infrastructures design

Chapter 8

Introduction

In the previous chapters, we have described the SMARTify approach to design data infrastructures for smart cities. One of the key objectives of the method we have developed is supporting the analysis and definition of business models, requirements and the design of a reference architecture for the data infrastructure. In this chapter, we provide examples of the applicability of our frameworks. The nature of decisions made when designing data infrastructures using our frameworks will fundamentally vary from cities to cities. This means that some business models' components or components of the closed loop value chain may not become relevant to decision makers, while others will find necessary to incorporate all SMARTify activities rigorously to both design new data infrastructure and take current infrastructures forward. We expect that different data infrastructure projects will use SMARTify in different ways simply because their problems and goals are different. As a result, the validation of the usefulness of the framework is subject to the context in which our approach is being applied.

The case studies presented in Part III of this book were chosen as the method of demonstration for the concepts and techniques described in this book for the following reasons:

- We were interested in demonstrating the suitability of the concepts and techniques in large-scale smart cities projects.
- We were fortunate to have applied and validated our frameworks in real-world projects.
- When performed in real situations, case studies provide practical and empirical evidence that a method is appropriate to solve a particular class of problems.

8.1 Case studies description

To demonstrate the usefulness of our frameworks in practice, we present two case studies:

Case Study A – Designing an Open Data Infrastructure: We provide a real case study involving the design of a data infrastructure for a large city in the UK. Our approach helped the decision makers to find a suitable open data platform among a set of platforms offered by leading open data suppliers. The reference

architecture designed using our frameworks provided a blueprint for decision makers to use by selecting technologies, data and their value network of
partners.

Case Study B – Advancing data infrastructures: The second real-world case
study involves the design and advancement of a large-scale smart cities data
infrastructure. We have conducted a comprehensive evaluation of SMARTify
method in the context of designing an alternative data infrastructure for London.
Though this case study used a combination of simulated and real-world data and
systems, we were able to get valuable insight on the effectiveness of the method.
This case study provided valuable information to allow the decision makers to
(i) find a suitable open data platform among the several platforms provided by
leading open data suppliers and (ii) outsource the development of a platform
based on the reference architecture designed using the SMARTify approach.

We emphasize that the objective of the requirements elicitation task presented
in the case studies of this book is not to be an exhaustive and prescriptive list of
all the requirements necessary for the data infrastructure in question. The task of
eliciting requirements of a data infrastructure is extraordinarily complex to model,
and it is even harder in a first attempt and considering the requirements and the
conflicting requirements emerging from all the domains of the business models.
It is even harder to state the high-level goals of the data infrastructure independently from the technologies to be deployed. Thus, the analysis of the selected case
studies involved interviewing stakeholders of smart cities and studying the documentation of the data infrastructure projects in order to provide the project knowledge
required for understanding how data infrastructure strategies for smart cities are being
designed.

Our objective here is to demonstrate how our business models framework provides directions on how to elicit business models-driven requirements and resolve
mismatches and conflicts that may pose risks to decision makers who are either
designing or procuring a data infrastructure for their cities. The case studies are not
intended to design with any precision the underlying physical infrastructure of platforms or architectures for the Internet of Things neither in the studied nor alternative
platforms. Furthermore, it does not claim to be exhaustive in terms of identifying
and analysing all the factors and requirements which may contribute to the efficient
management of city data. To an extent this is because what is studied is part of an
extremely large system-of-systems and there are most likely to be several extraneous
factors which have not been considered. Although we do not cite machine learning
algorithms, blockchain, IoT sensors' network, cloud computing platforms, it is to be
assumed that those technologies will be embedded in many of the several architectural
components and layers of the data infrastructures fulfilling the requirements elicited
for each one of them.

Finally, there does not exist one fits all solution, which means that the decision
for the best risk resolution relies on the judgment and experience of the providers of
the infrastructure. Once the business models and requirements have been analysed
and the potential reference architecture has been identified, the next step is to select

the best resolution available on the market, or design the solution in-house, or a combination of both.

8.2 Designing an Open Data Infrastructure

In this section, we describe a real case study concerning designing a data infrastructure for *Smart City Platform* project. The main objective of this case study is to demonstrate the use of each phase of our framework in detail to demonstrate its applicability to a real-world case and compare the approach with the current state of the art used by the practitioners in the smart cities industry team who were in charge of designing the Open Data Infrastructure. The case study was accomplished through information obtained from the relevant literature and confidential materials, a series of open interviews with directors, managers and consultants in the smart cities consulting firm. The presentation of the case study in the next section is structured according to the design steps of our approach.

Open Data Infrastructure

The UK Smart City would like to develop a data exchange, an analytical, and potentially commercially viable, open platform for citywide solutions. Smart City would now like to examine the business potential of building the Smart City Platform. This study examined the work done by Smart City and its proposal to the Technology Strategy Board competition for a *Future Cities Large-Scale Demonstrator* and sought to propose new, viable options and practical routes forward.

8.2.1 Business models outline

To obtain initial input for the Smart City Platform business model, we began by conducting interviews with the managers and designers of the company providing the data infrastructure, analysing the existing Open Data initiatives, the city data currently available and potential stakeholders, as well as literature materials which highlighted the open issues in data infrastructures. This analysis resulted in a first sketch of the business model, comparable to the result of Step 1 (Figure 6.1) and it is presented in the following sections.

SERVICE DOMAIN

The **value proposition** is a high-level goal or statement of benefits that are delivered by the platform to its external and internal stakeholders. The Smart City's proposal for *Future Cities Large-Scale Demonstrator* sets an overarching vision for the City:

> By the year 2026, Smart City and its citizens is to be able to enjoy a high quality of life (within environmental limits) and enriching cultural experiences become the overall outcome for the residents of the Smart City.

Table 8.1   Open Data Infrastructure value proposition

Scoring	Rationale
Value proposition	
Build a comprehensive city data platform that could attract further investment from the business community	
Rationale	The target delivered value of the Smart City Platform is to provide a data exchange, an analytical, and potentially commercially viable, open platform for citywide solutions

Table 8.2   Open Data Infrastructure – Intended Value (1)

Intended Value: Increase transparency and provide data and services the community needs
Value can be achieved through the transparency of data, ease of access to data, better segmentation and targeting of customer needs, and the deployment of better products. During the designing phase of the platform, it must be taken into consideration the requirements of providing open data, the data quality expected from users and the different levels of concerns with regard to data collection, integration, enrichment, distribution and reuse.

Concept	Rationale
Requirements	City data environment should be provided as open data
Stakeholders	Internal Data Providers
Tariff	Free
Resources	Business models, value network, information requirements

The value proposition of the Smart City 'Open Data Infrastructure' is shown in Table 8.1.

Three **Intended Value(s)** (sub-goals) were derived from the platform value proposition, and they refer to the value and opportunities the platform is intending to be offered. These three intended values are translated into functional requirements for the technology architecture (technology domain) and into requirements for the design of the value network (organization domain), and are presented in Tables 8.2–8.4. In these tables, we first define the **Intended Value** and its rationale and justification. What follows are the definitions of four concepts:

1. The key requirements that will enable the intended value to be translated into delivered value;
2. The stakeholders that influence the implementation and realization of the intended value;

Table 8.3   Open Data Infrastructure – Intended Value (2)

Intended Value: Increase user engagement with open data strategies
Economies of scale do not exist in most cities. Cities want to develop bespoke systems; it may be possible to identify the core datasets that feed into most services using the 80–20 rule (i.e. 80% of data is likely common). In order to get users to take up the platform, the open data supply chain needs to be efficiently managed and maintained. Therefore, a well-managed data supply chain has the potential to facilitate unlocking value from data by providing the right data at the right time and in the desired format.

Concept	Rationale
Requirements	City data should be available, have high quality and be trustable
Stakeholders	General consumers of open data
Tariff	Free
Resources	Business models, information requirements, data supply chain, system architecture, system processes

Table 8.4   Open Data Infrastructure – Intended Value (3)

Intended Value: Improve the efficiency of existing systems through the use of integrated open data.
Saving can be achieved through the deployment of better systems, driving efficiencies, reducing congestion and generating time savings, thereby creating productivity benefits as demonstrated by TfL journey planner generating benefits in the ratio of 58:1. Employers could be attracted to a city which provides a good transport system to its citizens. In the domain research, it became clear that many different technologies will rely upon the open data provided in this platform. Hence, it is of utmost importance to maintain good quality, standardized and interoperable datasets.

Concept	Rationale
Requirements	Data should be available in good quality, be standardized and interoperable
Stakeholders	General consumers of open data
Tariff	Free
Resources	Business models, value network, information requirements, data supply chain, system architecture, system processes

3. The tariff – if any – the end-users will have to bear in order to use the deliverables of the intended value;
4. The resources necessary to enable the realization and delivery of the intended value.

The Smart City Platform **delivered value** must meet the requirements originated from the datasets nature, the existing and future technologies and the value network of stakeholders. The UK Smart City had recently begun to attract foreign direct

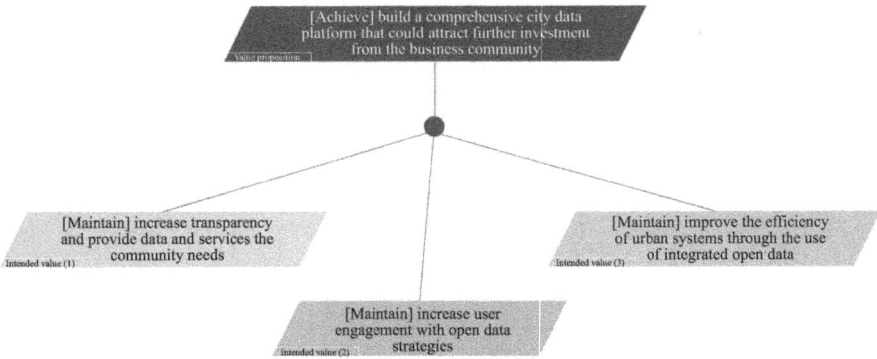

Figure 8.1   Value proposition and intended values of the Smart City Open Data Infrastructure

investment, and multi-nationals are more likely to respond to a local authority that actively supports the emergence of new private sector models and that behaves in a way that demonstrates their commitment to innovation. Applying a charge to open data will not help develop products and services expected to be offered via this open data portal. The open data market is still at an early stage of development and the market segment who would benefit from using open data are extremely price sensitive. It is believed that charging for data will result in a huge resistance from the very community that they need to have onside, that is, the developer and innovator community who will see this as an attempt for the open market to subsidize the public task of data collection.

The data infrastructure high-level goal (value proposition) and the three intended values (sub-goals) were defined to allow us to understand the high-level capabilities of the data infrastructure. This was the first step towards eliciting the requirements the platform must fulfil. Figure 8.1 illustrates the value proposition and intended value shown as high-level goal/sub-goals of the Open Data Infrastructure.

The **end-users** of the Open Data Infrastructure are – besides the Platform Owner – Data Consumers, Data Providers and Data Facilitators. In Tables 8.4–8.6, we present each end-user of the data infrastructure, their **previous experience** (if known), the context in which they use the data infrastructure, their respective market segment. Figure 8.2 illustrates the usage scenario for five different users of the data infrastructure. The platform provider is shown in this image under the ID number 1 alongside the other four stakeholders.

The **Data Consumer** is a user who may obtain open data from the platform. This user can contribute data through their use of city infrastructure, mobile technologies and information services. They also benefit from the use of these facilities by gaining improved access to city services, employment, friends and family.

The **Data Facilitator** is a user who provides data services outside the platform. Data facilitators are application developers and data regulators. They play an essential role in making sense of, and creating value out of, city data by developing innovative

Table 8.5   Open Data Infrastructure – Data Provider

Stakeholder: Data Provider

Concept	Rationale
Context of use	Data to fact
	City data providers will need to receive authorization to have access to services for data publications. Data may be manually uploaded or submitted via APIs. They will also need to have access to services and functions for updating, maintaining and accessing both data and metadata, as well as tracking the usage of resources by end users.
Market segment	Private sector organizations, city councils, international organizations, services providers
User ID (Figure 8.2)	2
Previous experience	Difficulty in publishing data
Expected value	• Easily publish data and metadata, check the quality of datasets, define data license • Manipulate data and associated metadata, update datasets, perform format conversions • Maintain the provenance, quality, trustworthiness and legal aspects of the data

Table 8.6   Open Data Infrastructure – Data Consumer

Stakeholder: Data Consumer

Concept	Rationale
Context of use	Data to fact
	City data consumers will need to access and use the city data residing in the Smart City Platform. End-users will be able to search metadata and full text within datasets (when available), and obtain city data in open formats readily available to both humans and machines such as CSV, XML, JSON. Some end-users may require different access rights to city data
Market segment	Open data users, including both national and international users (humans and machines)
User ID (Figure 8.2)	4 and 5
Previous experience	Difficult to find and use open data
Expected value	• Access to open data • Easily discover open data • Consume data that is human understandable • Obtain access to data infrastructure while preserving my identity and privacy

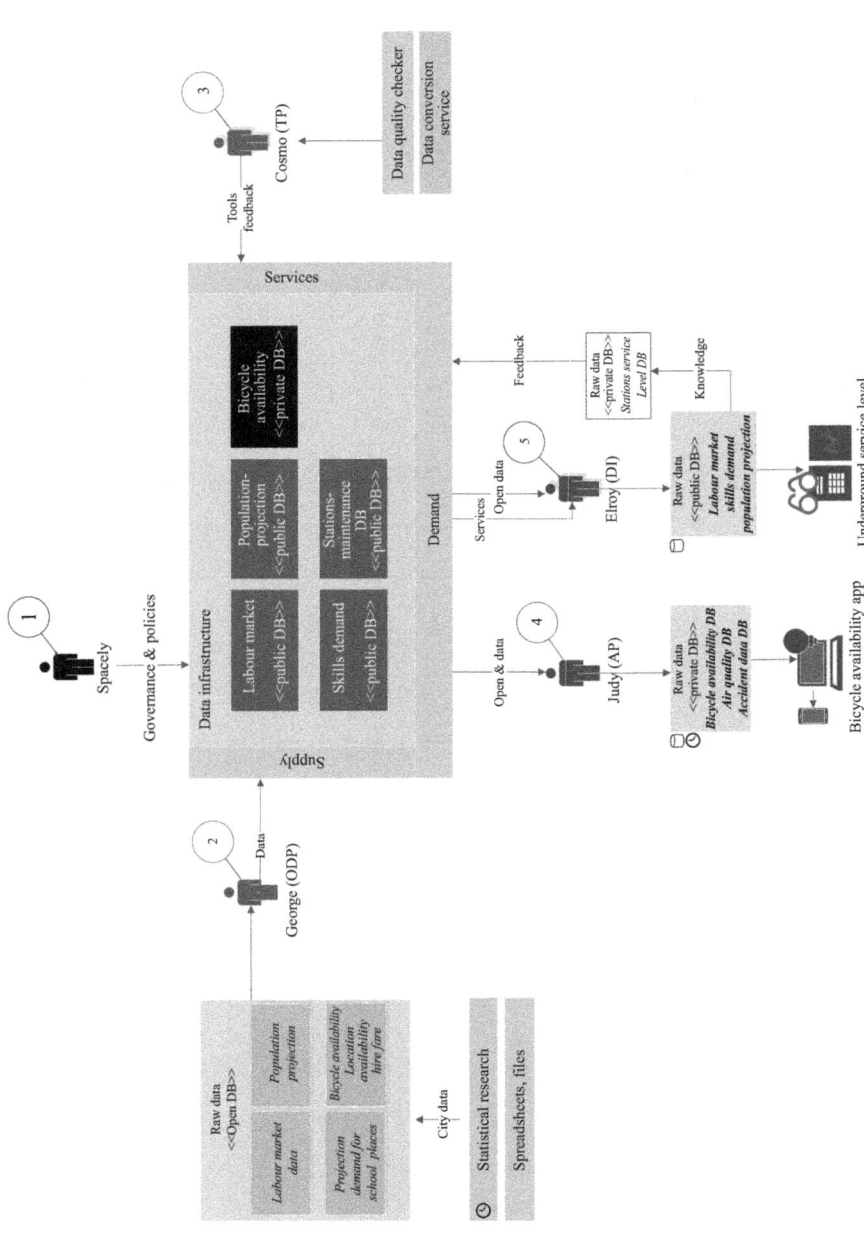

Figure 8.2 Open Data Infrastructure's stakeholders and their usage scenarios

Table 8.7   Open Data Infrastructure – Data Facilitator

Stakeholder: Data Facilitator	
Concept	**Rationale**
Context of use	Data to interface, data to service
Market segment	Data regulators (e.g. standardization bodies) and Application Developers
User ID (Figure 8.2)	3
Previous experience	Offer applications for data manipulation, standards and policies recommendations to Open Data Strategies

apps. Citizens often make informed decisions based on city data through mobile and web applications (Table 8.7).

The Smart City Platform should provide the appropriate mechanisms to receive city data from authorized data providers. Such mechanisms should allow data to be manually uploaded or submitted automatically via APIs. The data providers with whom the Smart City Platform negotiates submission agreements are the providers of open data. The Smart City Platform must provide data providers with specifications on the content, quality and format of acceptable data, and the terms and conditions for data publication. Once data is submitted to the Smart City Platform, it must undergo several reviews, including format compliance, metadata minimum requirement agreement, bias, quality and anticipated content and data formatting. When data is successfully submitted (either via APIs or manual upload), it will be processed/prepared for persistence into the Smart City Platform's elastic cloud database. The Smart City Platform must confirm the publication of city data to their providers and request them to resubmit data in the case of errors resulting from the submission.

The publication of city data by data providers may represent a legal transfer of custody for the data in the urban platform and may require that special access controls are placed on the contents so that the data providers can manage and update data. City data management should include services and functions for updating, maintaining and accessing both data and metadata, as well as tracking the usage of resources by users. Ideally, the owners of the resources should be the only authorized user to manage resources, and other authorized users can track the usage of the resources in the Smart City Platform. Data usage tracking may include performing queries on the data management data to generate result sets and producing reports from these result sets.

The Smart City Platform should execute its services within an expected fairness and performance profile, as well as handle on-demand increased processing volumes of data and service requests. However, it is very difficult to have clear performance characteristics due to the known demand for city data and services, and the real volumes of data to be managed in the platform. As part of the services provision, the infrastructure should prioritize service and data requests, monitor service level agreements, analyse the performance of the platform over time and scale up or scale out as necessary.

The concept of trust should be taken into account during the entire specification of the Smart City Data Infrastructure. In some cases, trust is associated with the reliability of data and their providers, whereas in other cases, it is associated with the security and privacy of the technology that was deployed. Trust may affect the reputation of the Smart City Platform besides its dissemination and maturity on the market. Hence, the platform will need to incorporate mechanisms to ensure the trust of users in the reliability, integrity and ability of the functional behaviour of the platform and should be addressed in the technology domain.

Ensuring users privacy is protected positively influences user's experience, acceptance and continuous use of the platform. Besides other factors, the reputation of the platform depends on how well user's information is secure and preserved. The providers of the Smart City Platform must check whether extensibility technical requirements impact on trust, avoid unauthorized access to implicit information (e.g. location) and verify the impact of security, trust and scalability requirements trade-offs on privacy.

8.2.2   Service design requirements

In summary, the service delivered by the data infrastructure should enable users to easily discover/publish, access and consume high-quality open data. In essence, the intended value is that users can have access to high-quality open data. Tables 8.8 and 8.9 present the data infrastructure functional and non-functional requirements based

Table 8.8   Service design functional requirements

Requirement ID	Rationale
FREQ.1	Allow data publishers to register to submit data for publication
FREQ.2	Store terms of agreements, and use them to monitor/review/process data submissions
FREQ.3	Able to add and edit terms of agreement, based on access of the level of user
FREQ.4	Monitor and manage data publications
FREQ.5	Provide mechanisms for static data publication
FREQ.6	Provide mechanisms for real-time data publication
FREQ.7	Enable data providers to manage their resources
FREQ.8	Maintain temporal information about the data
FREQ.9	Accept content in numerous file types/formats
FREQ.10	Verify the validity of the submission based on submitter, expected format, data quality, and completeness
FREQ.11	Promptly request data resubmission if any errors occur
FREQ.12	Store and track versions of data
FREQ.13	Create and maintain link/connections between versions of data and metadata
FREQ.14	Provide to users information about the provenance and legal aspects of the data
FREQ.15	Enable users to access terms and conditions to use the services of the platform

Table 8.9 Service design quality requirements

Class	Type	Requirement ID	Stakeholders' concern
Security	Run time	NFREQ.1	Provide secure access to resources at all times
	Run time	NFREQ.2	Ensure the integrity of the system from malicious services
	Run time	NFREQ.3	Provide trusted and secure communication and information management
Privacy	Run time	NFREQ.4	Keep users' access-control rights/ policies secured
	Run time	NFREQ.5	Provide privacy protection for users interacting with the platform
	Run time	NFREQ.6	Provide communication confidentiality
Availability	Run time	NFREQ.7	Guarantee infrastructure availability

on the service domain analysis and stakeholders concerns. In Figure 8.3, we illustrate an example of the mapping of services requirements in the city data value chain.

TECHNOLOGY DESIGN

Smart City Platform should be deployed upon a cost-effective approach which seeks to scale up as demand is generated, commensurate with resources available. The recommended approach is to start small and work with scalable technology and launch a low-cost platform that can be developed over time. Due to financial constraints, the initial investment in Smart City Platform should be kept to a minimum; however, the platform should be specified from the outset to be capable of handling the provision of new services and handling transactions in future. We suggest adopting an open-source platform already available in the community that meets the following specifications.

Backbone Infrastructure should be resilient and always available and scalable. **Interfaces** and **APIs** enable users to have access to the platform services. They should be able to invoke the platform Web services using Web service standard APIs (application programming interfaces). The owners of the Smart City Platform should adopt the concept of microservices and sharding to enable optimal performance and graceful degradation in case of failure.

Data infrastructure availability is another important aspect to be considered in the design of the Smart City Platform. The platform should be fully operational when required and effectively cope with failures that could affect system availability. For this, the owners of the platform will need to identify suitable backup and disaster recovery solution, adopt fault-tolerant hardware and rework the architecture as necessary to ensure high availability of services and resources. Standard interfaces should allow data consumers to communicate and obtain data from the data infrastructure, and deal with legacy systems which will allow for adaption and the next generation of systems. The Smart City Platform shall be evolvable and accommodate future requirements such as the commercialization of data and data manipulation services.

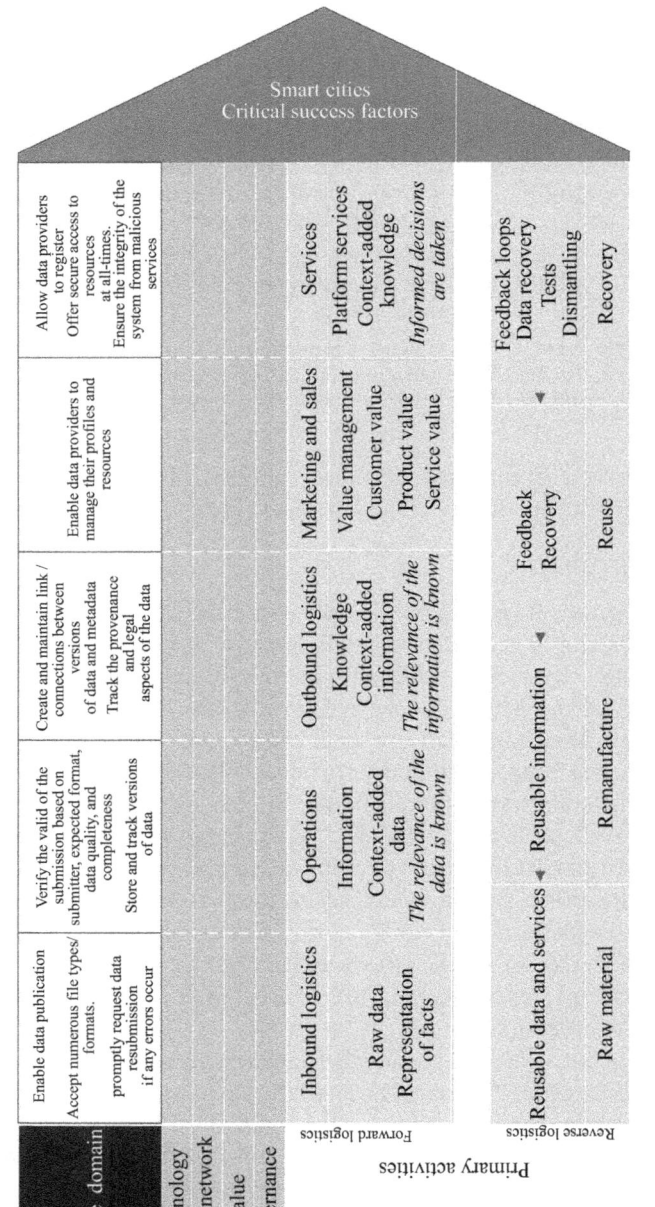

Figure 8.3 Mapping services requirements on the Data Infrastructure Reference Architecture

The **Service Platforms** for the data infrastructure should consist of a city data management component that offers flexible means for data publication and maintenance services, including via the web interface, rich API and custom spreadsheet importers. When data is successfully submitted to the Smart City Platform (either via APIs or manual upload), it is processed/prepared for storage into the data management platform. The data management platform includes the generation of unique identifiers to the datasets, enrichment of data with ontologies (when applicable) and standards to ensure that the data conforms to the platform data formatting, security and regulations. Current Metadata schemas (e.g. open data ontologies) should be reviewed to verify if it meets current needs for the data management platform. It is possible that the needs of the urban platform designed here may require new or additional schemas.

The **Applications** provided by the platform should be a user interface for searching data, an endpoint for users to subscribe and download data. The platform should also provide rich functionalities when it comes to harvest, extend, configure and version-control of **Data/Metadata**. It is very important that the data provided by the platform is machine readable and understandable. The platform shall expect different data architectural features, formats, high volume with different velocities, and vulnerability aspects such as sensitive personal information. Table 8.10 presents an extract of some datasets considered to be present in the infrastructure.

8.2.3 Technology design requirements

In this section, we present the functional and non-functional requirements derived from the technical domain. The requirements gathered from the technology domain focus on the physical and digital aspects of the platform. In essence, city data should be modelled using open standards, rich in metadata, of good quality, and discoverable. The platform should be scalable and enable different data collection, storage and retrieval speed, and support the storage of high data volume. Tables 8.11 and 8.12 present the general functional and non-functional requirements of the platform based on the technology domain analysis and stakeholders concerns.

VALUE NETWORK DESIGN

The main component in the value network design is the value network, which consists of several actors and their interactions. In the platform context, platform owners and their data providers are structural partners, users using and providing feedback in the platform are contributing partners and users providing regulations in the platform are supporting partners. Together they form the data-driven value network as illustrated in Figure 8.4 and presented in Table 8.13. In this figure, we illustrate the basic silos of the value network in the context of a knowledge value chain. In the first block of the diagram, the representation of facts are raw data gathered from city data sources (e.g. live streams, open data) which when is added with creates information, that is, it is possible to know the meaning of the data. Once the information is created it can be turned into knowledge which means the relevance of the information is known. For this, in the data operations silo data is contextualized, organized and stored.

Table 8.10 Open Data Infrastructure exemplary datasets

System	Dataset	Bytes/r	GB/day	Type	Architecture	Frequency	Sensitive
Parking	Parking location	–	49,438	Text data	Structured	Non-real time	Yes
Transport	Parking penalties data	–	49,438	Text data	Structured	Non-real time	Yes
Transport	Journeys	–	74	Text data	Structured	Non-real time	Yes
Transport	Railway stations	–	5,113	Text data	Structured	Non-real time	No
Transport	Road accident and safety	–	5,113	Text data	Structured	Non-real time	Yes
Transport	Bus stop signs	180	798	Text data	Structured Near	Real time	No
Transport	Buses tracking	200	31,311	Sensor data	Semi-structured	Real time	Yes
Transport	Taxi GPS traces	83	15,730	Sensor data	Semi-structured	Real time	Yes
Transport	Traffic disruption	307	1	Text data	Structured	Near real time	Yes
Transport	Variable message sign	180	7	Text data	Structured	Near real time	No
Transport	Traffic cameras	25,600	160,051	Image/data	Multi-structured	Near real time	Yes
Health	Cancer dataset	–	49,438	Text data	Structured	Non-real time	Yes
Health	Hospital complaints	–	49,438	Text data	Structured	Non-real time	Yes
Health	Death/mortality	–	78,102	Text data	Structured	Non-real time	Yes

Table 8.11   Technology design functional requirements

Requirement ID	Rationale
FREQ.16	Keep data and metadata secured
FREQ.17	Record a minimal set of identifying information/metadata concerning data publication submission
FREQ.18	Convert data to accepted file formats
FREQ.19	Keep sensitive information secured and accessible only to authorized users
FREQ.20	Keep user's personal information protected
FREQ.21	Keep city data and metadata secured
FREQ.22	Enable privacy-preserving mechanisms associated to data
FREQ.23	Model data in accordance with defined standards
FREQ.24	Have the ability to search and display metadata, preferably in a user-conformable, human-readable display as well as in its native format for machine harvesting and manipulation
FREQ.25	Control access to data in the repository based on multiple permission levels. These permission levels determine the create/edit/read/delete privileges granted users
FREQ.26	Hold access rights and conditions of use for each data and its related metadata
FREQ.27	Maintain the integrity of the database which contains both metadata and system information
FREQ.28	Provide internal validation such as referential integrity of the contents of the database
FREQ.29	Create and maintain schema definitions required to support data management functions
FREQ.30	Monitor and ensure that data and metadata are not corrupted during transfers
FREQ.31	Provide disaster recovery capabilities including data backup, off-site data storage, data recovery, etc.
FREQ.32	Refresh/replace data without service interruption, and update corresponding metadata as appropriate
FREQ.33	Ensure that any associated unique identifiers of the updated data are not altered
FREQ.34	Monitor functionality of the entire repository
FREQ.35	Maintains integrity of system configuration
FREQ.36	Provide standard and open APIs and query engines for users to discover data
FREQ.37	Enable users to consume open data anonymously
FREQ.38	Provide a modular based architecture which relies on stable and well-defined interfaces to ensure interoperability between the platform, services and the applications provided by the platform
FREQ.39	Provide multi-purposed and network intelligent interfaces

The knowledge can then become informed decisions at the end of the chain – outbound logistics – as data can be queried and discovered.

Platform providers are actors who participate in the design, development and sponsoring the platform technology. Data providers and data facilitators are niche players that provide data and standards recommendation. Data regulator is an entity that alongside the platform owner establishes standardization of procedures, data generation and interaction. The regulators of the platform work to assure best practices/recommendations, interoperability and production, aimed to efficiency and costs

Table 8.12   Technology design quality requirements

Class	Type	Requirement ID	Stakeholders' concern
Scalability	Run time	NFREQ.9	Support different service level agreements (SLA)
	Run time	NFREQ.10	Scale with regard to data volume
	Run time	NFREQ.11	Scale with regard to the number of users
Usability	Run time	NFREQ.12	Provide easy to use and informative interfaces
Availability	Run time	NFREQ.13	Ensure network availability
	Run time	NFREQ.14	Perform self-healing
Evolvability	Non-run time	NFREQ.15	Be extensible for future technologies and services
Security	Run time	NFREQ.16	Have security mechanisms to protect data transmission
	Run time	NFREQ.17	Perform to detect threats at runtime

reduction, and have the ability to set policy and regulations around data collection and data sharing, invest in digital infrastructure, collect data from citizens and city activities, and use that data to inform urban planning, design and operation. Data consumers use the platform to consume data and services.

The high-level goals of the stakeholders of the data infrastructure demonstrate that users want to consume enriched high-quality open city data, and application developers want to create valuable new businesses through integrated data. At the same time, data regulators want to guarantee such users have access to agreed standards, and that open data is protected from misuse and the data ownership is secured. Application Developers wants to obtain rich data in machine-readable formats in a timely manner.

Application developers can register in the platform and request access to the platform APIs. They provide valid registration details and wait for registration confirmation. Platform Providers may authorize or not application developers to use the platform's APIs. Application developers must formally agree with API and data usage agreement with the Smart City Platform. This agreement defines terms of the content, policies, regulations for service and data use and reuse. The Smart City Platform should proactively work with application developers to agree on the technical specifications of interfaces and platform openness level. Agreements between Platform and application developers may be renegotiated on a periodic or ad hoc basis.

VALUE DESIGN

On the basis of a successful launch of Smart City Platform being populated with useful open data, there is the potential for small businesses to begin to generate added value with the data through manipulation, visualization, connectivity and new applications and services. Smart City Platform will need to evolve over time and develop its own specializations; these may well deliver health benefits, travel time efficiencies and other new uses unique to the Smart City and with global applicability.

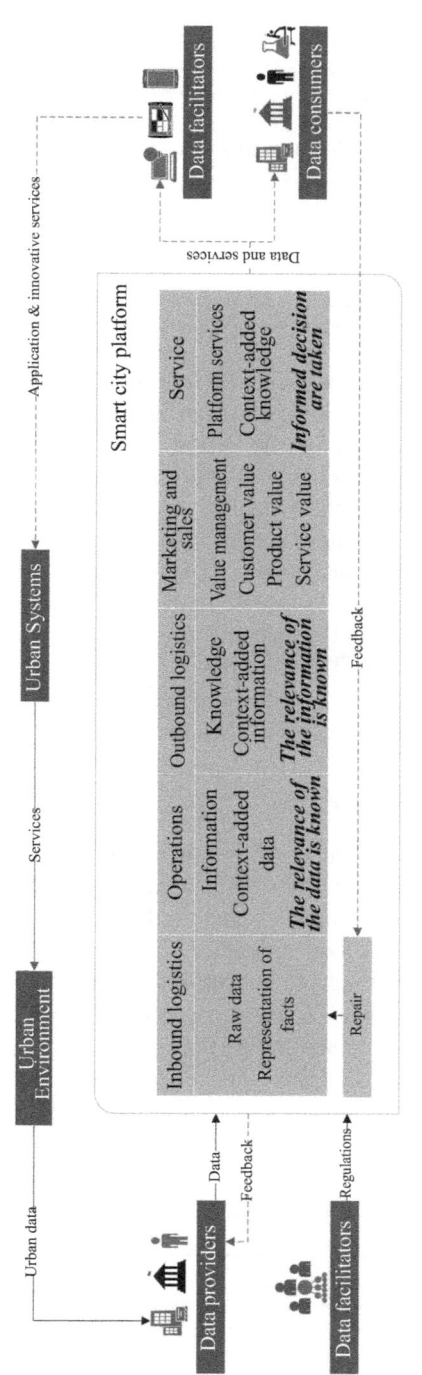

Figure 8.4   Smart City platform data-driven value network

Table 8.13   Data-driven value network members

Role	Category	Stakeholder	Resources
Keystone	Structural partner	Open Data Infrastructure provider	Provides the platform and coordinates the value network
Niche Players	Supporting partner	Data Publisher and Facilitator (Data Regulator)	Provides city data, ethical and standards recommendation
Niche players	Contributing partner	Data Consumer and Application Developer	Provides feedback and processes data for wider consumption

Furthermore, opening up data is not always free, and there are some potential costs associated with the production and presentation of open data that need to be considered and accounted for. Such matters raise the need to take special care when designing platforms for city data management. Such a platform must be designed in reasonable and affordable ways, which do not add unnecessary financial costs and potential loss of revenues. Smart City Platform does not envision the commercialization of data but acknowledges that there is a future possibility that it may happen. Hence, the only source of funding for the platform is to seek governmental funding.

To reduce the cost of the deployment and development of Smart City Platform, governmental and research council funding should be sought. The value and economic impact of the platform can be assessed by measuring user engagement and the services created through the use of the data provided by the platform.

8.2.4   Value design requirements

In this section, we present the functional and non-functional requirements derived from the value domain. In summary, the requirements gathered from the value domain focus on the financial aspects of risks and revenues. There is a substantial commitment and investment on the part of public agencies as they need to acquire new skills, train employees, get data specialists to prepare information to be released, purchase technologies and upgrade network infrastructure, which need to be accounted for. Pricing, Division of Investments and Risks, Division of Costs and Revenue are not possible to be captured at the R&D phase, hence such CDIs are not considered in this business model phase. Table 8.14 presents the general functional and requirements of the platform based on the value domain analysis.

Table 8.14   Value design functional requirements

Requirement ID	Rationale
FREQ.45	Capture statistical information about services and data usage
FREQ.46	Allow users to provide feedback on the usability and quality of data and services provided by the platform

GOVERNANCE DESIGN

The Open Data Infrastructure could act as a 'facilitator' bring together the silos of information for the sharing, analysing and distribution of information. Though it requires appropriate governance structure, data exchange, analytical and predictive capabilities. There is a need for new business and service models working around end-user satisfaction, not simply led by the supplier or public sector information.

The governance model of the platform should be centralized in order to give the platform owner and data regulators the power to strategically provide features which will be a differentiator from existing platforms which will attract data facilitators, knowledge creators, data providers and consumers to the platform. As in this case the platform owners retain strategic decision rights over a platform, they should retain the power to alter the rights and privileges of data providers and access to certain datasets. This gives a platform owner the flexibility to make changes to the degree of openness of the platform over time. However, we argue that platform owners should involve public sector information providers in the strategic decisions, because they are likely to be able to contribute two distinct types of knowledge that are needed by the platform owner and data regulators to make decisions.

An open and hybrid approach model for the delivery of city data seems to be appropriate for the data infrastructure. It supports the provision of information on demand as it shares features of a platform which provides real-time data through APIs and bulk download, and therefore can fulfil the requirements of data consumers who want to consume historical data and the data facilitators who want to create real-time applications.

8.2.5   Governance design requirements

In this section, we present the functional and non-functional requirements derived from the governance domain. Platform owners must shape and influence its ecosystem while respecting and involving complementors in the design choices. Table 8.15 presents the general functional requirements of the platform based on the governance domain analysis.

Table 8.15   Governance design functional requirements

Requirement ID	Rationale
FREQ.47	Monitor services usage
FREQ.48	Monitor data usage
FREQ.49	Provide platform owner with mechanisms to define the terms and conditions of platform usage
FREQ.50	Provide platform owner with mechanisms to authorize public sector information providers to publish data on the platform

8.2.6   Reference architecture model

Figure 8.5 illustrates a conceptual architecture view obtained from the business models requirements and the closed loop supply chain model. In this architecture, we present the three software layers which are responsible for inbound logistics, data operations and outbound logistics. For each of these layers, we have identified preliminary software components that will conduct distinct data activities inside the chain and serve as building blocks to unlocking value from the data.

In this architecture, the inbound logistics handle data publisher authentication, data connectivity and data quality checking. The authentication component provides the security of the system from a user access perspective. Data connectivity works as an abstraction layer to gather data from a myriad of data sources (spreadsheets, databases). Data quality checking ensures the data operations layer will not be fed by incomplete and poor quality data. Inside the data operations layer, the Data Wrapper component is responsible to semantically enrich the input data, model the data into a defined standard format and link the data to existing external open data sources. Data Manager stores the data and its related metadata. The components of the architecture and the respective requirements they fulfil are presented below.

LOGISTICS: FORWARD

Data Inbound Layer: At the bottom of the architecture resides the user authentication mechanism which authenticates users and allows fine-grain access control on the database. The authentication interfaces give access to the connectivity component which will gather the data and submit it to the data quality checker. After the verification of the completeness of the data, the data management platform generates unique identifiers to the datasets, enriches data with ontologies and standards to ensure that the data conforms to the platform data formatting, security and regulations. After completing these activities, the data is sent to the Data Operations Layer.
  - **Functional Requirements:** FREQ.1, FREQ.3, FREQ.5, FREQ.6, FREQ.7, FREQ.9, FREQ.10, FREQ.11, FREQ.17, FREQ.18, FREQ.19, FREQ.20, FREQ.21, FREQ.22, FREQ.23, FREQ.30
  - **Quality Requirements:** NFREQ.1, NFREQ.3, NFREQ.7, NFREQ.8, NFREQ.10, NFREQ.12, NFREQ.13, NFREQ.16

Data Operations Layer: This layer is responsible for enriching data with semantic and provenance annotations. The Wrapper component converts multiple formats and databases to well format standardized data according to the relevant predefined standards. Access to data is granted after identifying users and validating access against eligibility rules (e.g. open data, user has subscribed and has access to consume a piece of data). The data and its respective metadata, including ontology and provenance annotations, are stored by the Data Manager Component which is responsible for data persistence. When data retrieval request arrives in the data management platform, security mechanisms should validate user's rights to access the data. In the case users are granted access, the data management

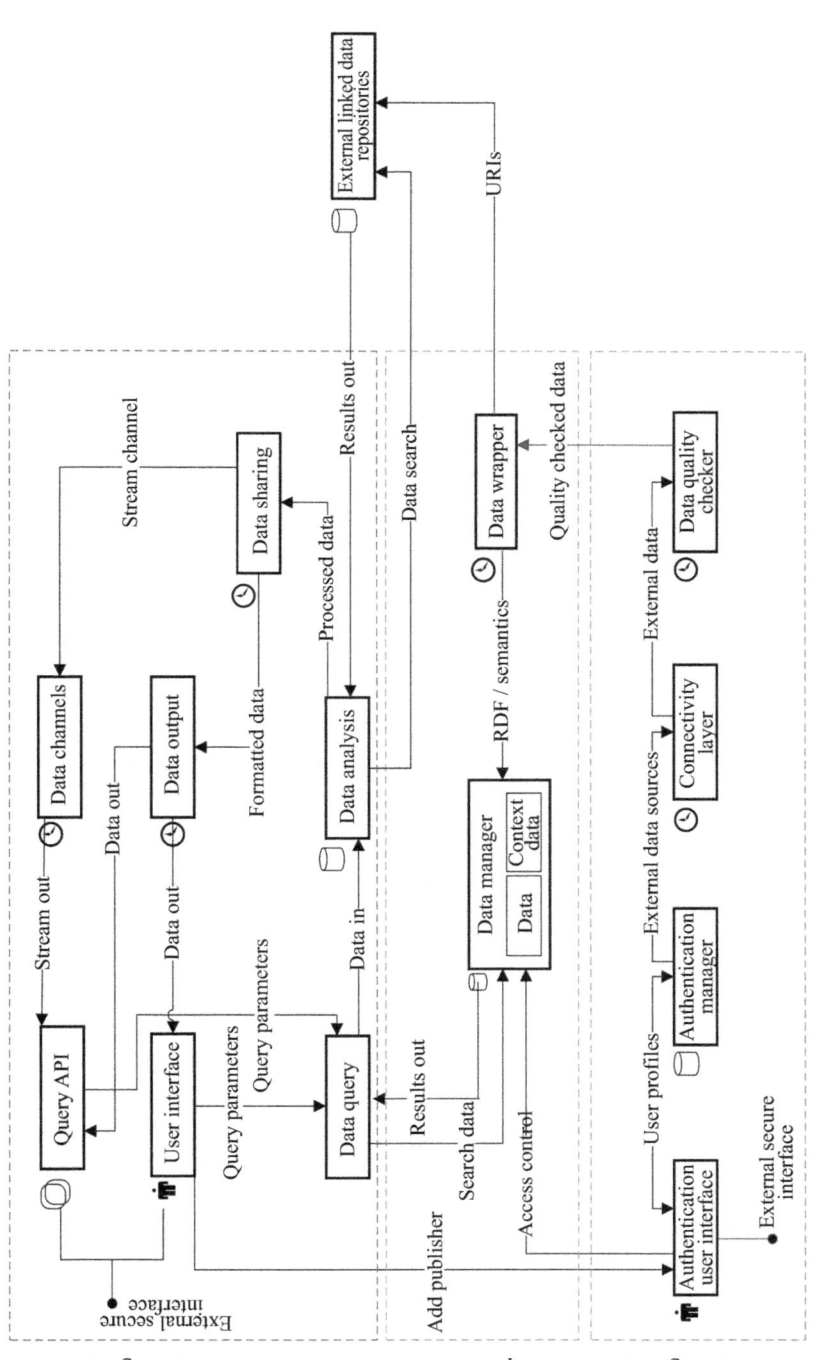

Figure 8.5   Reference architecture model of the Open Data Infrastructure

platform retrieves city data from data storage and moves a copy of the data to the Data Analysis component in the Outbound Logistics layer.

- **Functional Requirements:** FREQ.2, FREQ.4, FREQ.8, FREQ.12, FREQ.13, FREQ.24, FREQ.25, FREQ.26, FREQ.27, FREQ.28, FREQ.29, FREQ.31, FREQ.32, FREQ.33, FREQ.45, FREQ.47, FREQ.48, FREQ.49, FREQ.50
- **Quality Requirements:** NFREQ.2, NFREQ.3, NFREQ.5, NFREQ.7, NFREQ.9, NFREQ.10, NFREQ.11, NFREQ.13, NFREQ.14, NFREQ.15, NFREQ.16, NFREQ.17

LOGISTICS: FORWARD and REVERSE

Data Outbound Layer: The hybrid approach model was the platform strategy chosen. It shares features of a platform which provides real-time data through API's and bulk download. Data query component is responsible for answer data queries and requests from the user interface or query APIs. Data Analysis component received the data retrieved by the data management platform. If special processing is required, this component accesses data and applies the requested processes. The types of operations, which may be carried out, include sub-sampling in temporal or spatial dimensions, conversions between different data types or output formats, and other specialized processing (e.g. data visualization). Once it is finalized data will be sent to the appropriate delivery channels (e.g. API's, GUI). This layer also offers mechanisms to connect with external data sources on the web. Users are able to provide feedback on data and services via provided GUI. This layer is also responsible for managing the overall operation of the urban platform.

- **Functional Requirements:** FREQ.14, FREQ.15, FREQ.24, FREQ.36, FREQ.39, FREQ.40, FREQ.41, FREQ.42, FREQ.43, FREQ.44, FREQ.46
- **Quality Requirements:** NFREQ.4, NFREQ.6, NFREQ.12

Based on the blueprint architecture of the Open Data Infrastructure, the decision makers proceeded with the plan of outsourcing the Smart City Open Data Infrastructure to one of the leading open data suppliers called CKAN,[1] which provides *out-of-the-box* capabilities for Smart City Open Data Infrastructure v1.0, including basic data analytics, visualization tools, secure handling of data/different access rights, data processing and cloud storage. The CKAN architecture matches the business models and reference architecture designed using our frameworks approach.

The Open Data Infrastructure v1.0 also needs to take into consideration the maturity of the open data ecosystem and the supporting ICT in the Smart City at the time of implementation. Nonetheless, there is some supporting evidence from the suppliers' technology roadmap to show signs of developing features within their platform that are capable of meeting more sophisticated Smart City v2.0 requirements in the near

---

[1]http://ckan.org/

future. Although CKAN do not currently offer data commercialization capabilities, its open-source feature enables the platform to be integrated with third-party systems through Platform as a Service (PaaS) and/or Software as a Service (SaaS) to fulfil the sophisticated requirements such as the ability to commercialize data and allowing the providers of city data to join the value network of partners. By ensuring that the open-source platform is able to evolve, the Open Data Infrastructure v1.0 will not be restricted by proprietary platforms (such as Socrata[2]) and will be taken to the next level and serve the growing demand of data publishers willing to open up and commercialize data and data consumers looking for high-quality real-time data.

8.3   Taking smart cities forward

In this section, we describe the case study concerning designing a data infrastructure for London using a combination of real and simulated systems and datasets. The main objective of this case study is to assess the current data infrastructure in place, its evolution and current stage. We analysed the case study by applying each phase of the SMARTify approach in detail in order to demonstrate its applicability to a real-world/simulated case study and shall help eliminate incoherence. The case studies were accomplished through information obtained from relevant literature, a series of open interviews with directors and managers or the current infrastructure, and consultants in the field of smart cities and data infrastructures. The presentation of the case study in the next sections is structured according to the design steps of the SMARTify Approach.

8.3.1   London Data Infrastructure

Taking Open Data Infrastructures Forward

London is piloting a new infrastructure for city data management. The **value proposition** of this data infrastructure is to provide *ready access and delivery of all information and knowledge that unpins the decision-making process in smart cities.*

8.3.2   Dynamic business models analysis

In the first phase (R&D), the London Data Infrastructure[3] led to the creation of more than 200 apps, such as the Citymapper travel app, which has now been exported to some of the biggest cities in the world. Initially conceived as a *web portal* containing 50 datasets,[4] the GLA (Greater London Authority) proposed *'to release as much GLA*

---

[2]https://www.socrata.com/
[3]The data presented in this case study was collected through a series of interviews conducted in 2015–2016 from London's Government.
[4]http://www.opendatabasealliance.com/londondatastore.htm

Table 8.16   R&D phase – 2008

Driver	Rationale
Regulation	Increase transparency, improve and measure policy and the use of technology
Technology	Development of open data web portal
Market	Increased demand for open data
Feedback	Increased participation of application developers and users engagement
Niche Players	GLA Research group working on data strategies and technology adoption in London

Component	Rationale
Service	Offer open data through a web portal
Technology	Web portals with capabilities to host open data catalogues
Value Network	Task force for data opportunities involved mainly people from within GLA
Value	Offer free open data
Governance	Centrally controlled and participation was restricted by the GLA members

data as possible and to encourage other public agencies in London to do the same'.[5]
In this first phase, an important driver was found to be regulation and technology. At that time the *Capital Ambition* project was examining how policy was working across the London Boroughs, particularly regarding their use of new media and technology. The researchers noticed that governments were struggling to address issues in the provision of basic urban services (e.g. waste minimization and obesity, co-production of services) (Tables 8.16–8.19).[6]

They have concluded that public services should not only be efficient but communal and collective which includes citizens' participation. Open data was considered a vital component of that invitation to participation. This change in perception opened up a market opportunity for open data, and in phase two (Procurement) when the CKAN, an open-source platform, was chosen to host the London Datastore. CKAN can cope with over 2 million datasets, and with the Datastore, which can work with large amounts of data. It is entirely possible to use Access Control Lists (ACLs) to grant/deny access to a dataset to either an individual user or a group of users and provide a rich RESTful JSON API for querying and accessing dataset information and flexible means to enter/edit data of different formats in CKAN either via the web interface, rich API or custom spreadsheet importers.

In the third phase (Implementation/roll out), it offered the GLA the opportunity to experiment, first by providing a small collection of datasets on the web (50) to be used by app developers, citizens, city councils and any other interested party. In this early stage, the GLA was looking for and experimenting with an open data portal that could potentially succeed in the market. The liberalization of open data allowed both private and public sector, developers and entrepreneurs to optimize and/or create

[5] http://freelondonsdata.eventbrite.com/
[6] http://beyondtransparency.org/chapters/part-1/lessons-from-the-london-datastore/

Table 8.17   Procurement phase – 2008/09

Driver	Rationale
Regulation	Increase transparency, improve and measure policy and the use of technology
Technology	Open data portal with data publishing capability
Market	–
Feedback	Continuous engagement with developers offered valuable requirements to the solution
Niche players	GLA Research group working on data strategies and technology adoption in London, Platform providers and Developers

Component	Rationale
Service	Offer several categories of open dataset catalogues
Technology	CKAN platform adopted
Organization	Task force for data opportunities involved mainly people from within GLA
Value	Offer free open data
Governance	Centrally controlled by the GLA

Table 8.18   Implementation roll out – 2009/10

Driver	Rationale
Regulation	Increase transparency, improve and measure policy and the use of technology
Technology	Open data portal with data publishing capability
Market	–
Feedback	Continuous engagement with developers offered valuable requirements to the solution
Niche Players	GLA Research group working on data strategies and technology adoption in London, Platform providers and Developers

Component	Rationale
Service	Offer several categories of open datasets catalogues
Technology	CKAN platform adopted
Organization	Task force for data opportunities involved mainly people from within GLA
Value	Offer free open data
Governance	Centrally controlled by the GLA

new services, expand their markets inside and outside the UK (e.g. Citymapper travel app, Bike Share Map from CASA). This was a time when open data adoption had accelerated and appeared to have endless possibilities. In the UK specifically the conditions were optimal for a technological breakthrough in smart cities. Since its launch the Datastore has led to the creation of more than 200 apps, such as the Citymapper travel app, which has now been exported to some of the biggest cities in the world, and the Centre for Advanced Spatial Analysis' Bike Share Map, which shows bike hire usage and docking station availability in London and a range of cities globally.

Table 8.19   Market phase – 2010–Present

Driver	Rationale
Regulation	–
Technology	–
Market	Increased demand for open data and metadata provided through APIs
Feedback	Users frustrated for not having access to both static/real time high-quality data with standardized metadata attributes
Niche players	ODI, TfL, European Commission, Horizon 2020 projects, Research Institutes and group working on data strategies and technology across the world

Component	Rationale
Service	Static non-standardized open data offering
Technology	–
Organization	GLA, data providers, app developers and general data consumers
Value	New businesses created and increased user engagement
Governance	Centrally controlled by the GLA

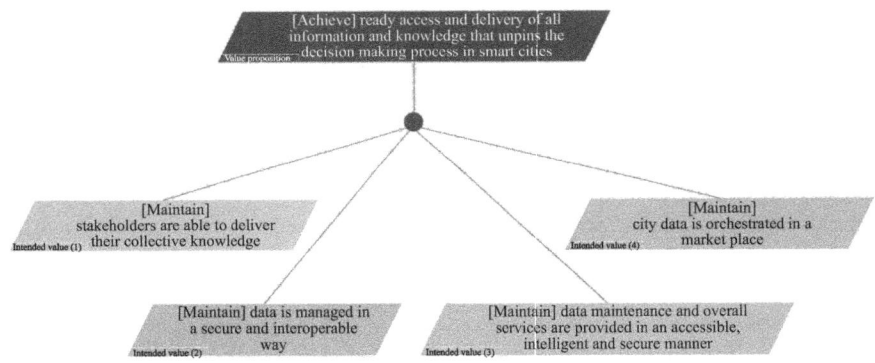

Figure 8.6   Value Proposition and Intended Values of London Data Infrastructure

As such, there is a link between the open data market and technology environment and the London Datastore business model. The latest sets include information on broadband connectivity levels so that businesses can choose where to locate to and to influence the market to improve services to homes and businesses. The site attracts 30,000 unique visitors a month, has led to the creation of more than 200 apps and hosts over 580 datasets. This indicates that there is a link between the open data market developments and the service component of the business model, more specifically the decision on targeting citizens and enabling them to have access to data that was previously not available. Tables 8.16–8.18 present the findings of the London Data Infrastructure case from phase 1 to 4, structured according to our approach (Figure 8.6).

More recently, with the help of feedback from users and new niche Players such as the Open Data Institute, the London Datastore has identified a promising market segment for open data. It has focused its attention on application developers with high entrepreneurial skills and a propensity to create ground-breaking innovation and applications, which have the potential to optimize the operation of urban services and the life of citizens. Besides, there is a whole new market demand for commercial high-quality data and a data infrastructure should need to fulfil the sophisticated requirements such as the ability to buy and sell data. However, the current infrastructure cannot be taken as a starting point since it does not have standardized metadata attributes, APIs, license scheme, advanced services for data manipulation, nor capabilities to enable the commercialization of datasets. The drawbacks of the existing infrastructure may limit the support that it can provide to the development of human, social, innovation and sustainable capabilities of London. A new business model has to be developed to guide the GLA on how to take the currently existing data infrastructure to the next level.

In the following sections, we present the business models analysis of this case study.

8.3.3 Business models outline

To obtain initial input for the London Data Infrastructure business model, we began by conducting interviews with the managers and designers of the existing London Data Infrastructure, analysing the existing Open Data initiatives, the city data economy in London, the city data currently available and potential stakeholders who could provide services and private/commercial data in the infrastructure, as well as literature materials which highlighted the open issues in data infrastructures. This analysis resulted in a first sketch of the business model, comparable to the result of Step 1 (Figure 6.1), and it is presented in the following sections.

SERVICE DOMAIN

Value proposition is a high-level goal or statement of benefits that are delivered by the data infrastructure to its external and internal stakeholders. The value proposition offered by the data infrastructure and its internal capabilities are shown in Table 8.23.

Three **Intended Value(s)** (sub-goals) were derived from the data infrastructure value proposition, and they refer to the value the infrastructure is intending to offer to the stakeholders of the London Data Infrastructure. These four intended values are translated into functional requirements for the technology architecture (technology domain) and into requirements for the design of the value network (organization domain). The four intended values (sub-goals) of the infrastructure are presented in Tables 8.20–8.24. In these tables, we first define the Intended Value and its rationale and justification, key requirement, related stakeholders, the tariff and resources necessary to enable the realization and delivery of the intended values. The target delivered value of the London Data Infrastructure is to provide an infrastructure that supports integrated and high-quality data, and where satisfied users are able to innovate and create valuable new businesses through integrated services. Hence, the delivered value must meet the requirements that come from data, technologies and stakeholders.

Table 8.20   London Data Infrastructure value proposition

Scoring	Rationale
Value Proposition	
Ready access and delivery of all information and knowledge that unpins the decision-making process in smart cities	
Rationale	City data stakeholders' needs to access data that is highly available, accurate, re-usable, valid, and that the source trustworthiness and the legal aspects of data consumption is reachable. Therefore, data on the infrastructure is published as open/private linked data. As a result, users are be able to integrate data from a variety of sectors and distribute it across different domains, value chains and stakeholders, thereby supporting intelligent decisions in the urban environment and the creation of valuable new businesses through integrated services. Therefore, a data infrastructure should deliver and incentive the development of advance features that will facilitate the collection, management and discovery of data.

Table 8.21   London Data Infrastructure – Intended Value (1)

Intended Value: Stakeholders of smart cities are able to deliver their collective knowledge
During the designing phase of the data infrastructure it must be taken into consideration the conflicting requirements of the stakeholders with regard to both private and open data, their strategic position in the infrastructure, and the different levels of concerns with regard to data collection, integration, enrichment, distribution and principles of data recycling and re-use.

Concept	Rationale
Requirements	City data environment should combine commercial, government, public and citizen data
Stakeholders	Internal and external data providers
Tariff	Free
Resources	Business models, value network, information requirements

Table 8.22   London Data Infrastructure – Intended Value (2)

Intended Value: Data is managed in a secure and interoperable way
In the domain research, it became clear that many different technologies and data are currently available in the urban environment. In order to get stakeholders to take up the infrastructure, the current and especially future complexity of all this complex data supply chain needs to be efficiently managed and maintained. Therefore, a well-managed data supply chain has the potential to facilitate unlocking value from data by providing the right data at the right time and in the desired format.

Concept	Rationale
Requirements	City data should be available, interoperable, secure and trustable
Stakeholders	Internal and external data regulators, publishers and consumers
Tariff	Free
Resources	Business models, information requirements, data supply chain, system architecture, system processes

Table 8.23   London Data Infrastructure – Intended Value (3)

Intended Value: Data maintenance and overall services are provided in an accessible, intelligent and secure manner

In the domain research it became clear that many different technologies and data are currently available in the urban environment. In order to get stakeholders to take up the data infrastructure, the current and especially future complexity of all this complex data supply chain needs to be efficiently managed and maintained. Therefore, a well-managed data supply chain has the potential to facilitate unlocking value from data by providing the right data at the right time and in the desired format.

Concept	Rationale
Requirements	Data services should be available, extensible, easy-to-use, secure, trustable, preserve privacy
Stakeholders	Internal and external data and services' consumers
Tariff	Freemium and premium services
Resources	Business models, value network, information requirements, data supply chain, system architecture, system processes

---

Table 8.24   London Data Infrastructure – Intended Value (4)

Intended Value: City data is orchestrated in a marketplace

The London Data Infrastructure aims to deploy state of the art data management strategies alongside business models and financial arrangements, and policies and licensing agreements, to deliver a foundation for widespread exploitation of data in the long-term. In order to accomplish these objectives, the data infrastructure will allow the providers of city data to commercialize data based on policies and fair financial models defined in the data infrastructure. The commercialization of city data involves the function of managing, updating, maintaining and accessing commercial data/metadata, and its respective commercial transactions.

Concept	Rationale
Requirements	Data subscription and payment mechanisms should be available, easy-to-use, secure and trustable
Stakeholders	Commercial data providers
Tariff	Freemium and premium data
Resources	Business models, financial models, value network, information requirements, payment capabilities, data supply chain, system architecture

---

In the London Data Infrastructure, data and service providers are not charged to publish their data/services. On the other hand, data and services users can consume free and charged data and services. In the data infrastructure, the open data business models will be adopted [388]. Basically, data providers who are able to provide high-quality and rich datasets may charge the data consumption in exchange for the effort they put on creating and opening up such knowledge. It is up to the end-user to choose whether to consume, for instance, a free limited version of a dataset or pay a tariff to consume an extended version (e.g. additional metadata) of that dataset. The innovation here is that a more demanding user can opt to use a richer version of data whereas a user who just needs to access a small portion of a dataset may have free access to it.

The data infrastructure high-level goal (value proposition) and the four intended values (sub-goals) were defined to allow us to understand the high-level capabilities of the infrastructure. This was the first step towards eliciting the requirements the London Data Infrastructure must fulfil. The London Data Infrastructure will put in place public and private governance structures and a shared data infrastructure to deliver city data. Embracing the technology and non-technology components of data infrastructures will ensure that standards are adhered to, interoperability is guaranteed, smart governance is put in place, a strong value network of partners are created, and feedback loops are facilitated. Figure 8.7 illustrates the value proposition and intended value shown as high-level goal/sub-goals of the data infrastructure.

The **customers and end-users** of the data infrastructure are professionals and researchers from within/outside governmental agencies, interested citizens and applications/systems providers (e.g. cloud computing). The customers and end-users of the London Data Infrastructure are, besides the data infrastructure owner, Data Consumers, Data Publishers, Knowledge Creators and Data Facilitators. Customer and end-user roles coincide with regard to consumer services but differ to business services and are considered as different entities. In the consumer services scenario we define stakeholders consuming data and services in the data infrastructure as customers/end-users. In the business services scenario, we define stakeholders providing data or services as the customers of the data infrastructure.

In Tables 8.25–8.28, we present each customer/end-user of the data infrastructure, their **previous experience** based on findings from the literature, the context in which they use the data infrastructure, their respective market segment. Figure 8.7 illustrates the usage scenario for five different users of the data infrastructure. The platform provider (Platform Keystone) is shown in this image under the ID number 1 alongside two Structural Partners who are the providers of commercial models, and standards and regulations which also include data ethics and services fairness.

Although there are no existing alternatives to this data infrastructure, previous experience of users can be derived from some platforms that currently provide some similar services. For instance, the London Datastore is a portal that releases open data to interested people in the form of a bulk download. On the other hand, the TfL developer provides real-time and historical data to authorized software developers. Users must sign up and request permission to consume transport data that are provided by the platform. In the London Datastore the customer is a particular government agency that uses the platform, and end-users are the users that interact and use the services provided by the platform.

The London Data Infrastructure must provide services and functions to accept the publication of city data from authorized data providers (of both open and proprietary data) and prepare the contents for storage and management within the data infrastructure. Functions include receiving data, performing quality assurance on data,

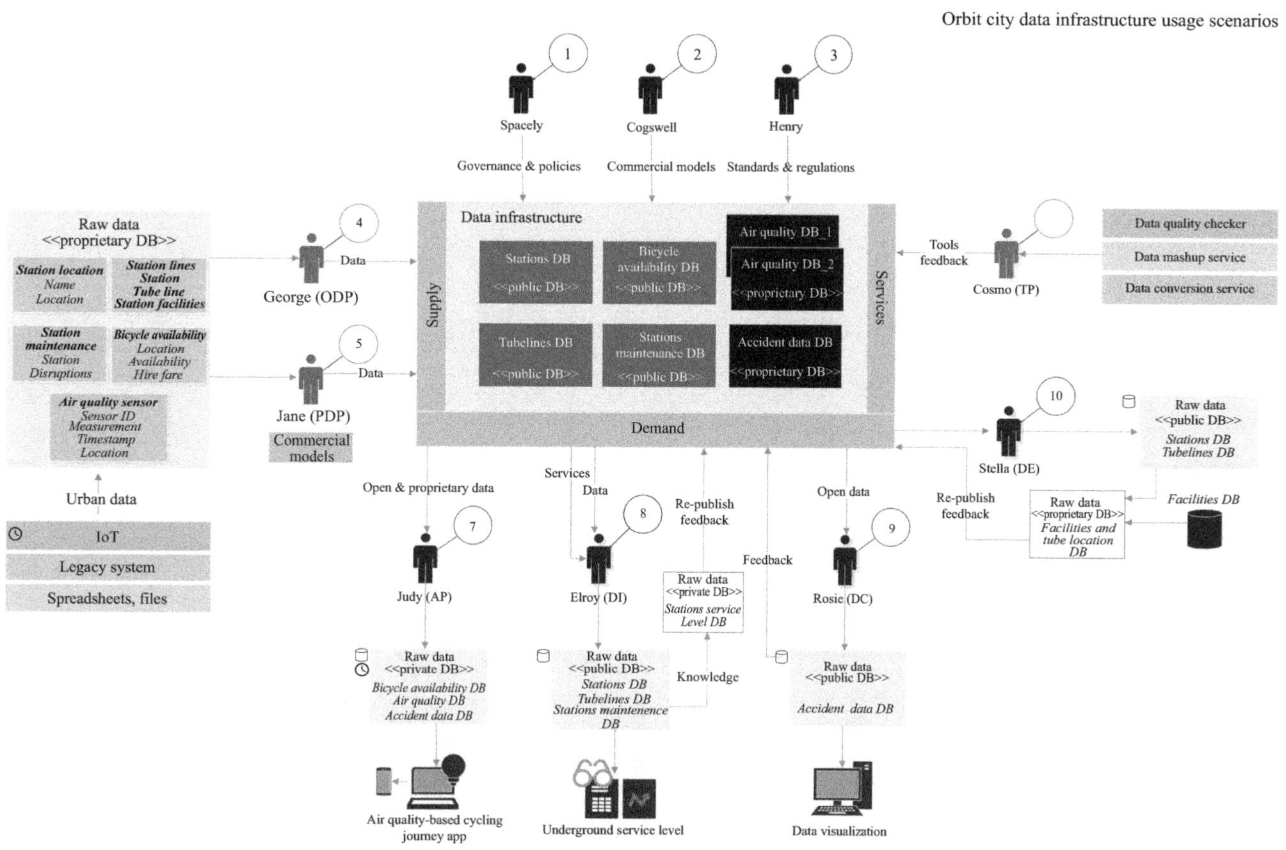

Orbit city data infrastructure usage scenarios

Figure 8.7 Open Data Infrastructure's stakeholders and their usage scenarios

Table 8.25   London Data Infrastructure – Data Publisher

Stakeholder: Data Publisher

Concept	Rationale
Context of use	Publication and management of data
Market segment	Professionals/researchers within non-government or government organizations
User ID (Figure 8.7)	4 and 5
Previous experience	Inability to guarantee data ownership and specify licenses attached to it
Expected value	• Easily publish open and private raw data in a machine/human-readable format using agreed standards • Provide data as a service and charge the release of private data • Define access-levels to the datasets to avoid data misuse • Guarantee my data is protected with agreed licenses, ownership and privacy agreements • Semantically enrich data to make them discoverable by external data repositories

Table 8.26   London Data Infrastructure – Knowledge Creator

Stakeholder: Knowledge Creator

Concept	Rationale
Context of use	Data to information, data to data
Market segment	Interested citizens, Professionals/researchers within non-government or government organizations, policymakers
User ID (Figure 8.7)	10
Previous experience	Inability to republish data
Expected value	• Easily share and re-publish high-level streams of knowledge • Integrate proprietary data with open data and publish on the data infrastructure • Track the provenance, quality, trustworthiness and legal aspects of the data

verifying data formatting and document standards, associating metadata information and coordinating updates to databases and resources management. Data Publishers can register in the data infrastructure and request approval to submit city data. They provide valid registration details (to be defined) and wait for registration confirmation. Data infrastructure (Platform) providers may authorize or not data publishers to offer both open and proprietary city data in the data infrastructure. Data submission agreement is a formal agreement between the data publishers and the London Data Infrastructure defining the terms of the content, standards, metadata creation and license agreement. The London Data Infrastructure will need to proactively work with

Table 8.27 London Data Infrastructure – Data Consumer

Stakeholder: Data Consumer

Concept	Rationale
Context of use	Data to fact
Market segment	Interested citizens, professionals/researchers within non-government or government organizations, policymakers
User ID (Figure 8.7)	7, 8 and 9
Previous experience	Difficult to find data, no understanding of licenses and provenance of data
Expected value	<ul><li>Have access to the domain knowledge of the data</li><li>Easily discover high-quality data and metadata in agreed standards</li><li>Consume data that is both human and machine readable/ understandable data</li><li>Obtain access to data infrastructure while preserving my identity and privacy</li><li>Track the provenance, quality, trustworthiness and legal aspects of the data</li><li>Be able to provide feedback on the quality and usability of the service and data</li><li>Download and stream data in formats of their choice</li></ul>

Table 8.28 London Data Infrastructure – Data Facilitator

Stakeholder: Data Facilitator

Concept	Rationale
Context of use	Data to interface, data to service
Market segment	Innovators, Tool Providers, Infrastructure Providers
User ID (Figure 8.7)	6 and 7
Previous experience	Inability to re-use data
Expected value	<ul><li>Access data through secure and available APIS</li><li>Access machine-readable data specifications</li><li>Provide applications to citizens and create profit</li><li>Provide applications and services at the application service level of the platform</li></ul>

data publishers to agree on the content, quality and format of city data. Agreements between the data infrastructure provider and data publishers may be renegotiated on a periodic or ad hoc basis.

Data publishers should receive a confirmation of receipt of data publication, which may include a request to resubmit data in the case of errors resulting from the submission. The publication of city data in the London Data Infrastructure may represent a legal transfer of custody for the data in the infrastructure and may require that special access controls be placed on the contents. The publishers of city data will require access to data management services and functions for updating, maintaining and accessing both data and metadata, as well as tracking the usage of resources by users. Ideally, the owners of the resources should be the only authorized user to manage resources, and other authorized users can track the usage of the resources in the data infrastructure. The infrastructure must provide a database update response indicating the status of the update, avoid update errors to be propagated in the data infrastructure and should keep an audit trail of all actions to enable rollback. Data usage tracking includes performing queries on the data management data to generate result sets and producing reports from these result sets.

Ultimately, the data infrastructure must ensure that its customers and users are able to publish, consume and commercialize data, as well as deploy and manage services all in a secure and privacy protected manner.

8.3.4   Service domain requirements

In summary, the London Data Infrastructure aims to deliver services to enable the stakeholders of city data to easily discover/publish, access, manage, purchase, consume/re-publish high-quality city data and deploy engaging new services. The intended value of this infrastructure is to provide a *custom-built experience* for all its users. Tables 8.29 and 8.30 present the London Data Infrastructure functional and non-functional requirements based on the service domain analysis and stakeholders' concerns.

TECHNOLOGY DESIGN

The London Data Infrastructure should be designed in a way it ensures that data is collected and sustained in accordance with well-established standards, managed in a robust manner so that it can handle high-level supply and demand of data which can be distributed and reused across different value chains, systems and stakeholders. The ability to managing the supply and demand of city data while providing enough security and accessibility to enable the city-wide exploitation of data is one of many keys to the success of the infrastructure.

The infrastructure **technical architecture** should be simple enough to be comprehensible at least at a high level of abstraction, that is, it should be conceptually decomposable into its major subsystems, the infrastructure's functionality reused by many services and external applications should be identifiable, and interactions between the infrastructure and services, data providers and data consumers should be well defined and explicit. The London Data Infrastructure shall be modular and rely on stable and well-defined interfaces to ensure interoperability between the platform, services and the applications developed by data facilitators. The infrastructure **technical architecture** is composed of the following technological components: applications, common data service platform, access networks, backbone infrastructure, interfaces and open APIs. Below, we provide details for each of these components.

Table 8.29   Service design functional requirements

Requirement ID	Rationale
FREQ.1	Allow data publishers (humans and machines) to register to submit open, proprietary and commercial data for publication
FREQ.2	Tracks data publication agreements between Data and Data Infrastructure Providers
FREQ.3	Store terms of agreements, and use them to monitor/review/process data submissions
FREQ.4	Able to add and edit terms of agreement, based on access of the level of user
FREQ.5	Support sensory data collection
FREQ.6	Monitor and manage data publications
FREQ.7	Allow authenticated users from across different organizations to publish city data
FREQ.8	Provide authorization mechanisms sensors to publish city data
FREQ.9	Enable the semantic description of connected devices
FREQ.10	Gather data from authenticated and authorized devices
FREQ.11	Validate automatically the successful transfer of the data
FREQ.12	Perform virus checking on data
FREQ.13	Provide mechanisms for static data publication
FREQ.14	Provide mechanisms for real-time data publication
FREQ.15	Enable the publication of metadata
FREQ.16	Maintain temporal information about the data
FREQ.17	Accept content in numerous file types/formats
FREQ.18	Verify the validity of the submission based on submitter, expected format, data quality and completeness
FREQ.19	Provide an interface for data providers to certify that the data and services they are providing has been verified against bias and unfairness
FREQ.20	Keep sufficient technical metadata to assure functionality (e.g. viewing and display) to ensure accessibility and reusability
FREQ.21	Allow publishers to display and perform manual/visual quality control assurance via a user-friendly GUI
FREQ.22	Store a minimal set of identifying information/metadata concerning data submission and publication
FREQ.23	Promptly request data resubmission if any errors occur
FREQ.24	Store and track versions of data
FREQ.25	Create and maintain link/connections between versions of data and metadata
FREQ.26	Enable users to access terms and conditions to use the data provided in the data infrastructure
FREQ.27	Enable city data to be consumed
FREQ.28	Enable city data to be streamed and downloaded
FREQ.29	Provide users information about the provenance and legal aspects of the data
FREQ.30	Enable data consumers to access terms and conditions to use the services of the data infrastructure
FREQ.31	Enable users to stream data through standardized and open APIs
FREQ.32	Store and tracks versions of data

(Continues)

Table 8.29   (Continued)

Requirement ID	Rationale
FREQ.33	Enable data providers to maintain and repair data and metadata
FREQ.34	Track data publication agreements between Data and Data Infrastructure Providers
FREQ.35	Store terms of agreements, and use them to monitor/review/process data submissions
FREQ.36	Able to add and edit terms of agreement, based on access of the level of user
FREQ.37	Manage and monitor data submission volumes and schedules
FREQ.38	Provide a user-friendly method of mapping non-standard metadata elements into approved standard elements which ensures data persistence, and allows a variety of access/distribution options

Table 8.30   Service design quality requirements

Class	Type	Requirement ID	Stakeholders' Concern
Security	Run time	NFREQ. 1	Provide secure access to resources at all times
	Run time	NFREQ. 2	Ensure the integrity of the system from malicious services
	Run time	NFREQ. 3	Provide trusted and secure communication and information management
Trust	Run time	NFREQ. 4	Provide privacy protection for users interacting with the data infrastructure
Privacy	Run time	NFREQ. 5	Keep users access-control rights/policies secured
	Run time	NFREQ. 6	Provide privacy protection for users interacting with the data infrastructure
	Run time	NFREQ. 7	Provide communication confidentiality
Availability	Run time	NFREQ. 8	Guarantee infrastructure availability
	Run time	NFREQ. 9	Be available for service when requested by end-users and applications

The **Backbone Infrastructure** refers to the medium- and long-range backbone network infrastructure which will provide the platform services for the data infrastructure. For the London Data Infrastructure we have considered Windows Azure as the Cloud Computing infrastructure (similar to theTfL platform) which is also used in the software simulation experiment of this case study. The **Interfaces** and **APIs** enable users to have access to the platform services and data facilitators to integrate their services in the platform. They should be able to invoke the platform Web services using Web service standard APIs (application programming interfaces). Standard interfaces should allow data consumers to communicate and obtain data from the data infrastructure, and service complementors to communicate and deploy services that augment the capabilities of the platform.

One defective service should not cause the entire ecosystem of the infrastructure to malfunction [240]. The key to such resilience is to ensure that services are weakly coupled with the infrastructure through interfaces that do not change over time. The **Service Platforms** for the data infrastructure should consist of a city data management component that supports graph data storage and manipulations, a city data subscription and payments mechanisms, users' data management (e.g. personal/context data). The **Applications** provided by the data infrastructure should be a user interface for searching data, an endpoint for users to subscribe and download data, and services for data quality checking and data augmentation. Changes in services should not require parallel tweaking in the data infrastructure; hence, the data infrastructure should be partitioned into standalone subsystems and then linking them using standardized interfaces. Rules and policies development will be needed regarding ingesting data and deploying data services into the data infrastructure, including which users/machines will be authorized to submit data for publication, the minimum requirements for data submitted by open and proprietary data publishers, and the removal of resources and services from the data infrastructure.

Another very important component of the technical domain is **Data/Metadata**. It is desirable that an infrastructure for city data management provides data in open formats (public specifications) that is both human and machine-readable and understandable. Data usability is a prime differentiator and it is a valuable competitive asset that increases efficiency, increases the service perceived value of the customer/end-users and drives profitability.

The London Data Infrastructure will require more expansive, robust and useful data encoding and conversion than what is available in the existing London Datastore. Data preservation policies should be developed to allow data to be stored in formats that can be migrated, be associated with metadata and ontologies to become both humans and machines readable and understandable, be monitored for data obsolescence and be migrated to systems environments as needed to ensure their continued availability. The city data found in the existing London Datastore may require special consideration concerning the type of formats and datasets that must be stored in the London Data Infrastructure. Current Metadata schemas (e.g. open data, sensory data ontologies) should be reviewed to see if it meets current needs for city data management. It is possible that the needs of the data infrastructure designed here may require new or additional schemas. Therefore, the system design and architecture should minimize fragmentation of city data in the data infrastructure and automate the extraction of descriptive and technical metadata. Table 8.31 presents an extract of some datasets considered to be present in the infrastructure.

Furthermore, once data is submitted for publication in the data infrastructure, it must undergo several reviews, including virus checking, format compliance, metadata minimum requirement agreement, quality and anticipated content and data formatting. The infrastructure will need to address the issues surrounding the 9 v's of city data. We expect different data architectural features, formats, high volume with different velocities and vulnerability aspects in the data provided by the London Data Infrastructure. To mitigate interoperability issues, the London Data Infrastructure will use

Table 8.31 London Data Infrastructure exemplary datasets

System	Dataset	Bytes/r	GB/day	Type	Architecture	Frequency	Sensitive
Metering and sensing	Energy metering	360	49,438	Sensor data	Semi-structured	Near real time	Yes
Metering & Sensing	Gas metering	360	49,438	Sensor data	Semi-structured	Near real time	Yes
Metering & Sensing	Buildings sensing	50	74	Sensor data	Semi-structured	Near real time	Yes
Metering & Sensing	Taxi sensing	50	74	Sensor data	Semi-structured	Near real time	Yes
Metering & Sensing	Smart lights	180	52	Sensor data	Semi-structured	Non- real time	Yes
Metering & Sensing	Water metering	260	78,552	Sensor data	Semi-structured	Near real time	Yes
Metering & Sensing	Water sensing	112	103	Sensor data	Semi-structured	Near real time	Yes
Transport	Bike hiring	50	1	Text data	Structured	Near real time	Yes
Transport	Bus stop signs	180	798	Text data	Structured	Near real time	No
Transport	Buses tracking	200	31,311	Sensor data	Semi-structured	Real time	Yes
Transport	Congestion charge	51,200	537	Image/data	Multi-structured	Near Real-Time	Yes
Transport	Oyster card	133	339	Text data	Structured	Near real time	Yes
Transport	Parking meter	50	92	Sensor data	Semi-structured	Near real time	Yes
Transport	Smart parking	102	350	Sensor data	Semi-structured	Near real time	No
Transport	Smart phones GPS	73	5,113	Sensor data	Semi-structured	real time	Yes
Transport	Taxi GPS traces	83	15,730	Sensor data	Semi-structured	Real time	Yes
Transport	Traffic cameras	25,600	160,051	Image/data	Multi-structured	Near real time	Yes
Transport	Traffic disruption	307	1	Text data	Structured	Near real time	Yes
Transport	Tube status	307	1	Text data	Structured	Near real time	Yes
Transport	Variable message sign	180	7	Text data	Structured	Near real time	No
Environment	Smart bins	27	20	Sensor data	Semi-structured	Non-real time	No
Environment	Smart environment	42	133	Sensor data	Semi-structured	Real time	Yes
Environment	Smart waste	73	2	Sensor data	Semi-structured	Non-real time	Yes

linked data and semantic web technologies to guarantee that data is represented in such a way that it is both human and machine readable and the relationships between data are understandable and discoverable. The data infrastructure will adopt well-established license agreements, fair commercial and subscription models to allow open and proprietary data providers to co-exist and cooperate in the data infrastructure (co-opetition model).

The data infrastructure will also include the ability to record all actions and decisions made concerning the publication of city data. The reasons for publication failure (e.g. missing metadata information, non-valid dataset) will be provided back to the city data publisher. In some cases, the data publisher can then resubmit corrected data and metadata information, while in other instances data publication refusal criteria should prevent the publisher from submitting the same dataset at a later time period (e.g. in cases of suspicious datasets – copyrights violation, viruses). Once data is successfully submitted (either via APIs or manual upload), it will be processed/prepared for storage into the data infrastructure's database.

The **technical functionality** refers to the functionality offered by the technological system and are often presented as a functional requirement list. To elicit the requirements for the technical domain of the data infrastructure, it is necessary to have a comprehensible understanding of the data, the services provided and the usage patterns of the data infrastructure's users.

8.3.5   *Technology design requirements*

In this section, we present the functional and non-functional requirements derived from the technical domain. We first start by providing a discussion of the following technological architecture components: data, the backbone infrastructure and access networks. In summary, the requirements gathered from the technology domain focus on the physical and digital aspects of the data infrastructure. In essence, city data should be modelled using public standards, rich in metadata, of good quality, discoverable and linkable. The data infrastructure should be scalable and enable different data collection, storage and retrieval speed, and support the storage of high data volume. Tables 8.32 and 8.33 presents the general functional and non-functional requirements of the data infrastructure based on the technology domain analysis and stakeholders concerns.

VALUE NETWORK DESIGN

Data infrastructure providers (or Platform Providers/Key Stone) are actors who participate in the design, development and sponsoring the data infrastructure. They have the most restrictive access level in the data infrastructure. This restrictive access-level grants the data infrastructure providers with ultimate rights to the system, including, but not limited to, infrastructure's management, development, deployments, assignment of access rights to data and services providers, policies and regulations, and license agreements. They should also be provided with access to services to follow up on civic engagement (e.g. feedback, request for city data) and on the provision of city data and services.

Table 8.32   Technology design functional requirements

Requirement ID	Rationale
FREQ.39	Model data using linked data modelling and ontologies
FREQ.40	Keep data and metadata secured
FREQ.41	Record a minimal set of identifying information/metadata concerning data publication submission
FREQ.42	Convert data to accepted file formats
FREQ.43	Model data in accordance with defined standards
FREQ.44	Support database-level provenance annotation
FREQ.45	Support data-level provenance annotation
FREQ.46	Enable data to be encrypted
FREQ.47	Provide services that enable users to manipulate city data
FREQ.48	Provide standard and open APIs and query engines for users to discover data
FREQ.49	Enable users to federate data from other open data endpoints
FREQ.50	Provide personalized services for registered users
FREQ.51	Have the ability to search and display metadata, preferably in a user-conformable, human-readable display as well as in its native format for machine harvesting and manipulation
FREQ.52	Keep sensitive information secured and accessible only to authorized users
FREQ.53	Provide built-in checks on the incoming metadata. Data not containing the minimally defined set of attributes should be returned to the publisher for metadata enhancement
FREQ.54	Keep user's personal information protected
FREQ.55	Keep city data and metadata secured
FREQ.56	Enable privacy-preserving mechanisms associated to data
FREQ.57	Provide access rights and conditions of use in machine readable and actionable format
FREQ.58	Provide granular access mechanisms to allow the identification of individual users, in order to maintain audit logs of actions performed by users
FREQ.59	Control access to data in the repository based on multiple permission levels. These permission levels determine the create/edit/read/delete privileges granted users
FREQ.60	Hold access rights and conditions of use for each data and its related metadata
FREQ.61	Maintain the integrity of the database which contains both metadata and system information
FREQ.62	Provide internal validation such as referential integrity of the contents of the database
FREQ.63	Create and maintain schema definitions required to support data management functions
FREQ.64	Monitor and ensure that data and metadata are not corrupted during transfers
FREQ.65	Provide statistically acceptable assurance that no components of the data are corrupted during any internal data transfer
FREQ.66	Perform routine and special data integrity checking for each dataset and generates error reports
FREQ.67	Audit submissions to ensure that they meet archive/repository standards
FREQ.68	Maintain configuration management of the system hardware and software
FREQ.69	Ensure data integrity for version upgrades and format migration

(Continues)

Table 8.32   (Continued)

Requirement ID	Rationale
FREQ.70	Provide disaster recovery capabilities including data backup, off-site data storage, data recovery, etc.
FREQ.71	Refresh/replace data without service interruption, and update corresponding metadata as appropriate
FREQ.72	Ensure that any associated unique identifiers of the updated data are not altered
FREQ.73	Audit system operations, quality of service and usage
FREQ.74	Provide quality of service information and database holdings inventory reports
FREQ.75	Monitor functionality of the entire repository
FREQ.76	Maintain integrity of system configuration
FREQ.77	Enable users to consume open data anonymously
FREQ.78	Provide a modular based architecture which relies on stable and well-defined interfaces to ensure interoperability between infrastructure services
FREQ.79	Provide multi-purposed and network intelligent interfaces
FREQ.80	Allow inventory, report on and migrate the contents of the data infrastructure
FREQ.81	Enable users to be authenticated

Table 8.33   Technology design quality requirements

Class	Requirement ID	Stakeholders' Concern
Scalability	NFREQ.10	Support different service level agreements (SLA)
	NFREQ.11	Scale with regard to data volume
	NFREQ.12	Scale with regard to the number of users
	NFREQ.13	Balance its load at runtime
Usability	NFREQ.14	Provide intelligent and easy-to-use interfaces
Supportability	NFREQ.15	Accommodate additional applications and technologies
Extensibility	NFREQ.16	Provide services in an interoperable manner
Reliability	NFREQ.17	Be available for service when requested by end-users and applications
Availability	NFREQ.18	Be able to perform self-healing
	NFREQ.19	Guarantee infrastructure availability
	NFREQ.20	Ensure network availability

Data providers and data facilitators are niche players that use core modules of the data infrastructure to provide data and services, and are responsible to maintain their resources in the data infrastructure. These partners require level access which enables them to carry out activities such as data publication, data quality verification, data manipulation, format conversion, data-access level permissions, financial models and licences definition. Furthermore, they require data maintenance access for reviewing or editing appropriate data and metadata in the data infrastructure. Data publishers should be granted permission to view, add or edit metadata without changing the data itself. They should be provided with access to feedback from users to investigate problem in their resources (e.g. missing data, inconsistent metadata), and statistical information about how their resources are used by users.

Data facilitators should be granted different access levels in the data infrastructure. Application developers should be granted with access to specialized APIs in order to obtain rich data in machine-readable formats and in a timely manner. Tools and services providers require access level which gives them rights to deploy services in the data infrastructure. They should be able to carry out activities such as data services deployment access (available to service providers adding new mechanisms or integrating new applications to the infrastructure), testing and validating integration, defining data-access level and tariff for service usage. Data services maintenance access is necessary for services providers reviewing, extending or editing applications in the data infrastructure. The providers of tools and services should be able to view services deployed and add to or edit access level and tariff without having to deploy the services again. They should be provided with access to feedback from users to investigate problem in their services (e.g. bugs, scalability issues), and statistical information about how their services are used by users.

Finally, knowledge creators and data consumers use the data infrastructure to obtain data and services, as well as provide feedback. City data consumers will need to access and use the city data residing in the data infrastructure. End-users will be able to search metadata and full text within datasets (when available), and obtain city data in open formats readily available to both humans and machines such as CSV, XML and JSON. Some end-users may require different access rights to city data. For instance, open data users may have special access granted to commercial data when purchase requirements and licenses are waived by the data provider.

Throughout this book, we have emphasized that the collaboration among all the members of the value network is required for the growth of a data infrastructure. Collaboration points in the value network can be identified by using the KAOS agents dependency diagram [183] which illustrates the relationships among actors in the data infrastructure environment as it exists before the data infrastructure is introduced. The relationships can be analysed in terms of opportunities and vulnerabilities. Figure 8.8 illustrates the stakeholders' (agents) dependency diagram of the London Data Infrastructure.

The main component in the organization domain is the value network, which consists of several actors and their interactions. In the data infrastructure context, users providing products (data) on the data infrastructure are structural partners, users providing knowledge and insights into the data infrastructure are contributing partners and users providing services in the data infrastructure are supporting partners. Together they form the data infrastructure value network which is illustrated in Figure 8.9.

In this figure, we illustrate the basic silos of the value network in the context of a data value chain. In the Inbound Logistics, raw data represents facts collected from the real world (e.g. data sources such as live streams, sensory data). When raw data reaches the Data Operations silo of the data supply chain, it is enriched with contextual

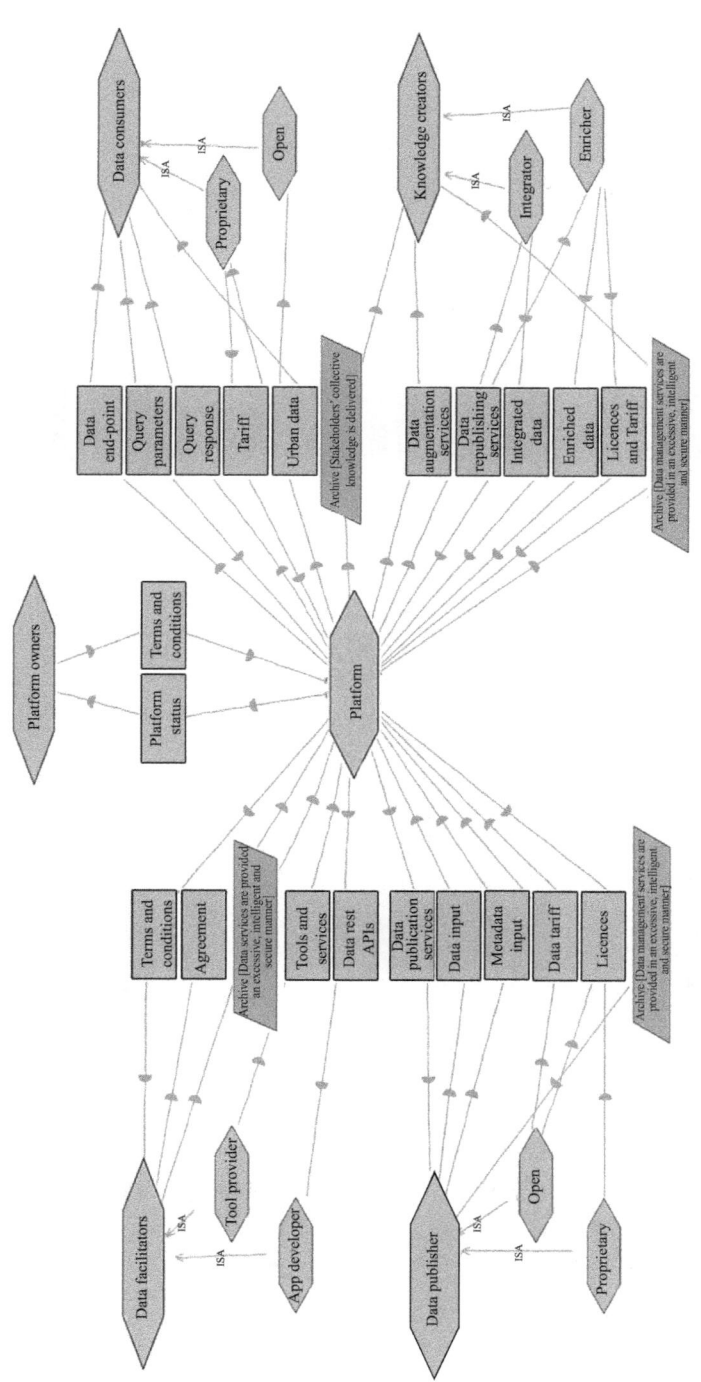

Figure 8.8   Stakeholders' Dependency Diagram

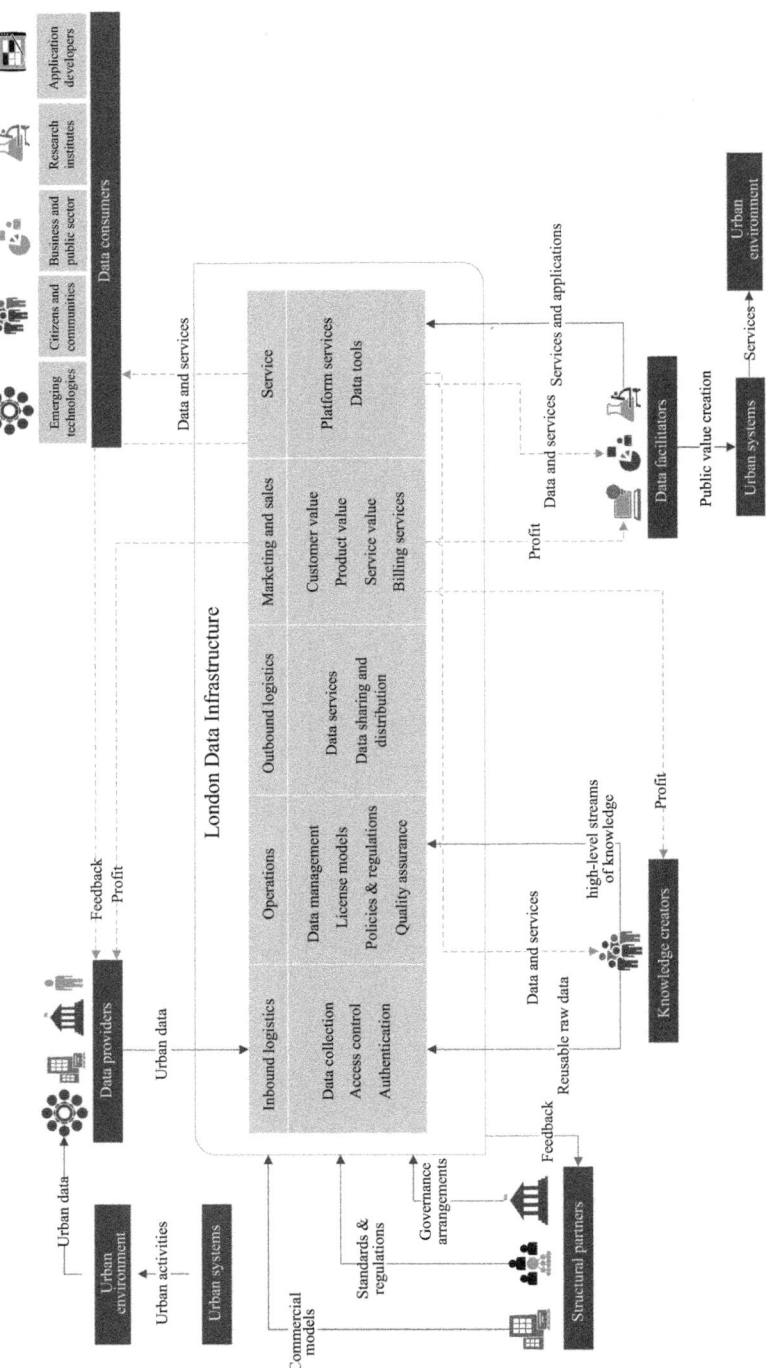

Figure 8.9   London Data Infrastructure Value Network

data and is transformed into information, that is, it is possible to comprehend its meaning. In the data operations silo the raw data is contextualized, organized, stored and linked to other data on the web. Once the information is properly stored and linked to other sources it is transformed into knowledge, that is, the relevance of the information is known. The knowledge can then become informed decisions at the end of the chain – data outbound – as data can be integrated, discovered, analysed and queried, and therefore users are able to apply knowledge in an appropriated way. The intelligence of a data infrastructure relies on how the various systems, value chains and stakeholders are interconnected and how the knowledge is passed through.

8.3.6   Value network design requirements

In this section, we present the functional and non-functional requirements derived from the organization domain. In summary, the requirements gathered from the organization domain focus on the financial, technological and services aspects of the data infrastructure. In essence, the stakeholders of the London Data Infrastructure have different requirements and expectations, and they should co-exist in such dynamic data infrastructure so that they can collaborate in delivering public value creation. Tables 8.34 and 8.35 present the general functional and non-functional requirements of the data infrastructure based on the organization domain analysis.

VALUE DESIGN

The London Data Infrastructure should be designed in reasonable and affordable ways, which do not add unnecessary financial costs and potential loss of resources. The providers of the data infrastructure will need to commit and invest in acquiring new skills, train employees, get data specialists to prepare information to be released, purchase technologies and upgrade network infrastructure.

With regard to the charge of data and services provided in the data infrastructure, users should be able to define the price/tariff of the services and data they are providing. Charging users to consume data may increase competitiveness on the data infrastructure, in which data providers will compete in providing the best quality data to end users. Data consumers may opt to consume data that has commercial costs or strictly consume open data that is free of charge. The data infrastructure must support both financial models.

The commercial exploitation of city data and their funding models are concepts that the London Data Infrastructure will need to address. The providers of the data infrastructure and their supporting partners must define license agreements and fair commercial and subscription models to allow interoperable open and commercial data as well as free and paid services to be exploited in the infrastructure.

The providers of city data can commercialize city data based on the policies and financial models defined in the data infrastructure. After publishing their data, publishers can define which data can be available as open data and which data should be available with the payment of a subscription fee. Once publishers define which data is to be commercially exploited, the data infrastructure will associate the data with their respective financial models and let it ready for subscription. City data

Table 8.34   Value network design functional requirements

Requirement ID	Rationale
FREQ.82	Enable data facilitators to access terms and conditions to provide services to the data infrastructure
FREQ.83	Enable data providers to manage their resources
FREQ.84	Provide stable and open interfaces to ensure interoperability between the services and the applications provided by services providers
FREQ.85	Provide multi-purposed and network intelligent interfaces to providers and consumers of services
FREQ.86	Provide service providers mechanisms to define the terms and conditions of data infrastructure services deployment
FREQ.87	Allow users to provide feedback on usability, and quality of services provided by the data infrastructure
FREQ.88	Enable agreements between data infrastructure providers and data facilitators to be edited/renegotiated on a periodic or ad-hoc basis
FREQ.89	Enable data facilitators to extend the services of the data infrastructure
FREQ.90	Enable users to subscribe to city data or services through the payment of a tariff
FREQ.91	Allow users to pay a tariff for using services
FREQ.92	Enable application of taxes on the commercialization of services and data
FREQ.93	Give data infrastructure providers access to relevant financial transactions on the data infrastructure
FREQ.94	Capture statistical information about services and data usage
FREQ.95	Utilize secure and reliable billing and payment management systems
FREQ.96	Enable data providers to manage the subscription models of their data
FREQ.97	Enable service providers to manage the commercial models of their services
FREQ.98	Enable city data and services to be commercialized

Table 8.35   Value network design quality requirements

Class	Requirement ID	Stakeholders' Concern
Extensibility and maintainability	NFREQ. 21	Be extensible and enable restructure
	NFREQ. 22	Be extensible for future technologies
Modifiability	NFREQ. 23	Be loosely coupled

consumer chooses which data to purchase and is redirected to a billing interface where the subscription payment is taken. The data infrastructure must provide an update response indicating the status of the payment. If successful, data is readily available to be consumed by humans and machines, otherwise the user can re-try the payment or cancel transaction. To enable the commercial exploitation of city data and services, the infrastructure will need to be flexible enough to accommodate different local, National and International data protection, licensing and commercialization regulations.

Commercializing city data also involves the function of managing the commercial data. It provides services and functions for updating, maintaining and accessing both data and its respective commercial transactions. Furthermore, it enables data providers to track the usage of commercial data by users. Ideally, the owners of the data should be the only authorized user to manage resources, and other authorized users can track the usage of the data in the data infrastructure. Data usage tracking includes performing queries on the data management data to generate result sets and producing reports from these result sets.

Data consumers should be provided with functions which enable them to manage their subscriptions and financial transactions on the data infrastructure. These functions include updating, maintaining and accessing financial transactions. For all these functions and services, the data infrastructure must provide a database update response indicating the status of the update, avoid update errors to be propagated in the data infrastructure and should keep an audit trail of all actions to enable rollback.

Initiatives such as converting government data to semantic web and linked data formats can be time consuming and therefore costly, similar to protecting partial pieces of data on datasets. An alternative solution is to invest on a long-term solution and design databases with the right of public access in mind (e.g. one-way encryption on databases). The issues and difficulties involved in modelling city data can be mitigated through the use of open-source applications that enable such a task, for instance, D2RQ[7] which converts relational databases into linked data. This would enable linked data to be released at lower costs, and as a consequence, increase its consumption by data consumers/data enablers/knowledge creators.

Figure 8.10 illustrates the Revenue and value sources of the London Data Infrastructure.

8.3.7   Value design requirements

In this section, we present the functional and non-functional requirements derived from the value domain. In summary, the requirements gathered from the value domain focus on the financial aspects of risks and revenues. Table 8.36 present the general functional requirements of the data infrastructure based on the value domain analysis.

GOVERNANCE DESIGN

The providers of the London Data Infrastructure must shape and influence its ecosystem, not to direct it, besides respecting the autonomy of complementors while also being able to integrate their varied contributions into a harmonious whole. The governance strategy of the infrastructure involves platform-based competition and focuses on the development of appealing features, driving innovation by complementors, and resolves the conflicting requirements among the actors who contribute to the platform by either providing data or services.

---

[7]http://d2rq.org/

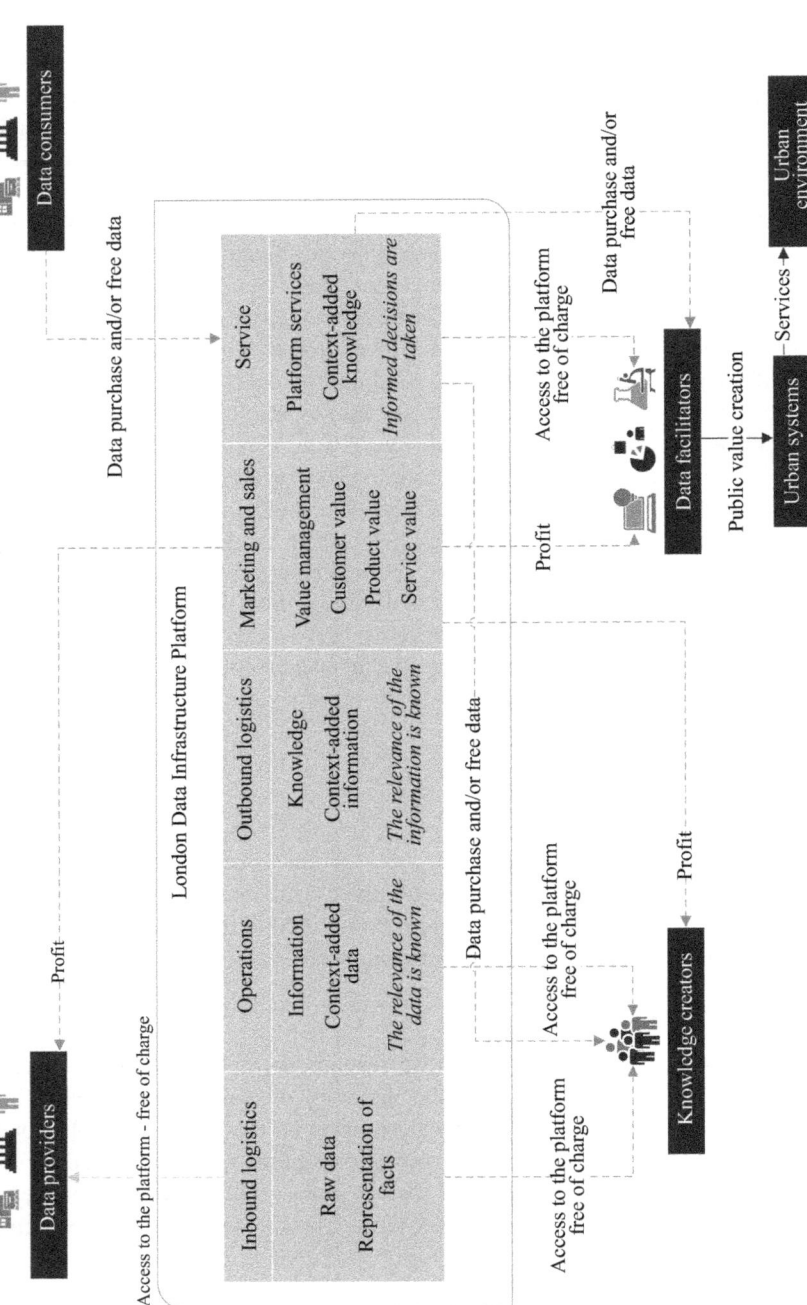

Figure 8.10 Revenue/value sources

Table 8.36   Value design quality requirements

Class	Requirement ID	Stakeholders' Concern
Robustness	NFREQ. 24	Handle faults and invalid inputs
Security	NFREQ. 25	Provide secure financial transactions

There are significant data gaps when it comes to open and commercial city data. This lack of data can, in some cases, lead to inertia with certain datasets not be released or conversely, undue attention being given to datasets that are unlikely to generate significant value but have a low cost of dissemination. There are a number of routes to addressing these data gap, which ranges from a detailed, regular audit of urban datasets to improve tracking of current usage. It further reinforces the importance of supporting feedback loops that enable service providers (e.g. data/knowledge providers) to understand the value customers perceive from using their data and services, and when necessary, to change their strategies so the users of city data and services are satisfied.

For the London Data Infrastructure, it makes sense to centralize platform strategic decision rights and give the platform owner the power to alter the rights and privileges of users and set contractual obligations and rules of participation in the value network. This gives the London Data Infrastructure's platform owner the flexibility to make changes to the degree of openness of the platform over time. However, we argue that platform owners should involve stakeholders in the strategic decisions, because they are likely to be able to contribute two distinct types of knowledge that are needed by the platform owner for making decisions. For instance, data facilitators are likely to better understand their own needs for their development work around a platform, while data consumers and app developers are closer to and represent the great pulse of emerging end users who will be driving the demand for city data.

Finally, an open and hybrid approach model for the delivery of city data seems to be appropriate for the data infrastructure. It supports the provision information on demand as it shares features of a platform which provides real time data through APIs and bulk download, and therefore can fulfil the requirements of the stakeholders of London's city data.

8.3.8   Governance domain requirements

In this section, we present the functional and non-functional requirements derived from the governance domain. Platform owners must shape and influence its ecosystem while respecting and involving complementors in the design choices. Table 8.37 presents the general functional requirements of the platform based on the governance domain analysis.

Table 8.37   Governance design functional requirements

Requirement ID	Rationale
FREQ.99	Monitor data usage
FREQ.100	Provide statistical information of users' feedback on service and data usage
FREQ.101	Provide data infrastructure providers with mechanisms to define the terms and conditions of platform usage
FREQ.102	Provide data infrastructure providers with mechanisms to define the terms and conditions of platform services
FREQ.103	Provide data infrastructure providers with mechanisms to define standards and protocols of the platform
FREQ.104	Provide data infrastructure providers with mechanisms to authorize platform complementors to provide services on the platform
FREQ.105	Monitor services usage

8.3.9   Reference architecture model

Figure 8.11 illustrates a conceptual architecture view obtained from business models requirements and the closed-loop supply chain model. In this architecture, we present the six software layers which are responsible for inbound logistics, data operations, outbound logistics, market and sale, services and reverse logistics services. For each of these layers, we have identified preliminary software components that will conduct distinct data activities inside the chain, and serve as building blocks to unlocking value from the data. Figure 8.12 presents a holistic view of the London Data Infrastructure architecture.

In this architecture, the inbound logistics handles user authentication, data connectivity and data quality checking. The Data Operations layers handle semantic annotation and provenance of the input data, and model data into Linked Data format and store it, as well as link the data to existing internal/external linked data sources. The Data Outbound layer deals with requests for city data and the preparation of data for consumption. The Data Service layer answers and processes users' requests for data via APIs or GUIs and monitors the quality of service of the data infrastructure. The Marketing and Sales layer handles data subscriptions and the financial transactions in the data infrastructure. Finally, the Recycling layer handles data maintenance and updates. The components of the architecture and the respective requirements they fulfil are presented in the following sections. Figure 6.13 illustrates a holistic view of the London Data Infrastructure and its respective data flows.

LOGISTICS: FORWARD and REVERSE

Data Inbound Layer: At the bottom of the architecture resides the user authentication mechanism which authenticates users, and allow fine-grain access control on the triple store. The authentication interfaces give access to the connectivity component which provides the appropriate mechanisms to receive city data from authorized data publishers. Data Publishers can register in the data infrastructure

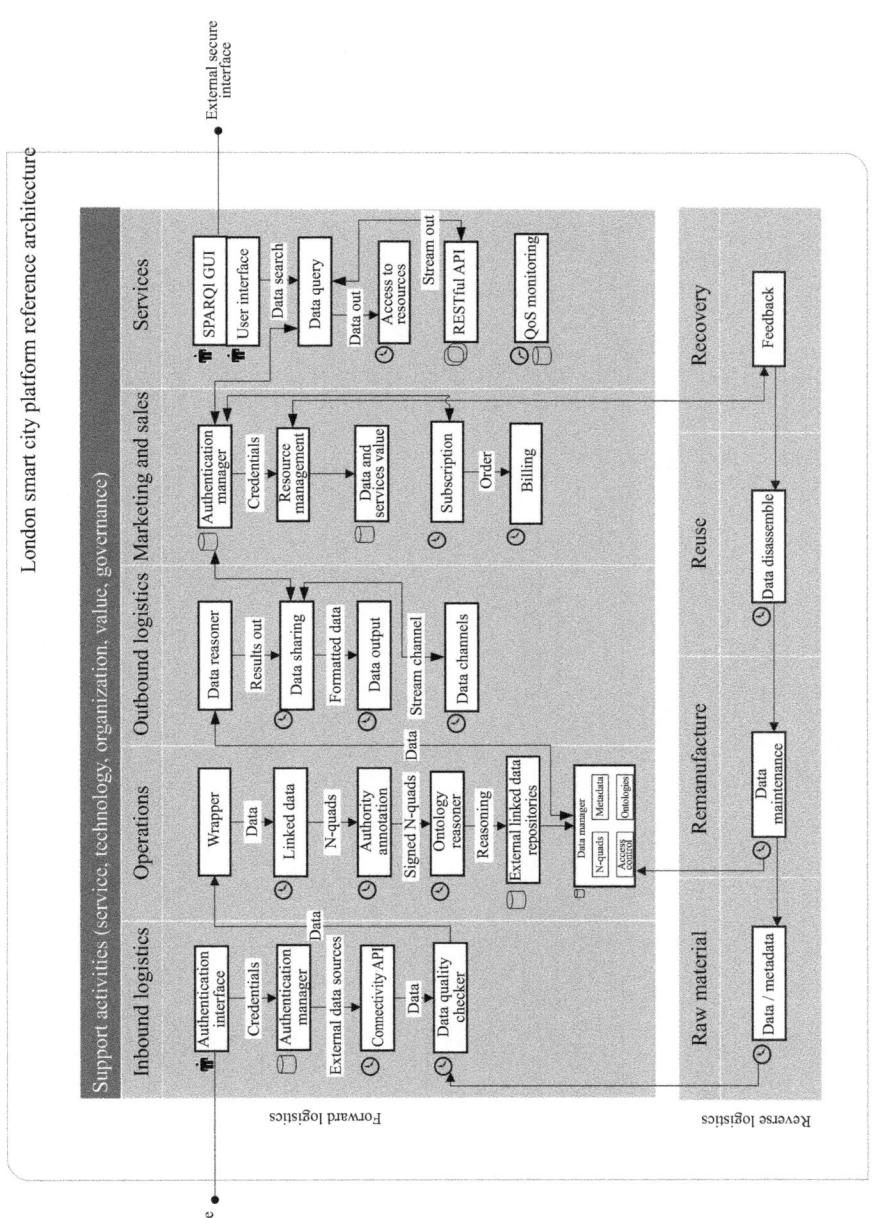

Figure 8.11   Reference Architecture Model overlapping the value chain framework

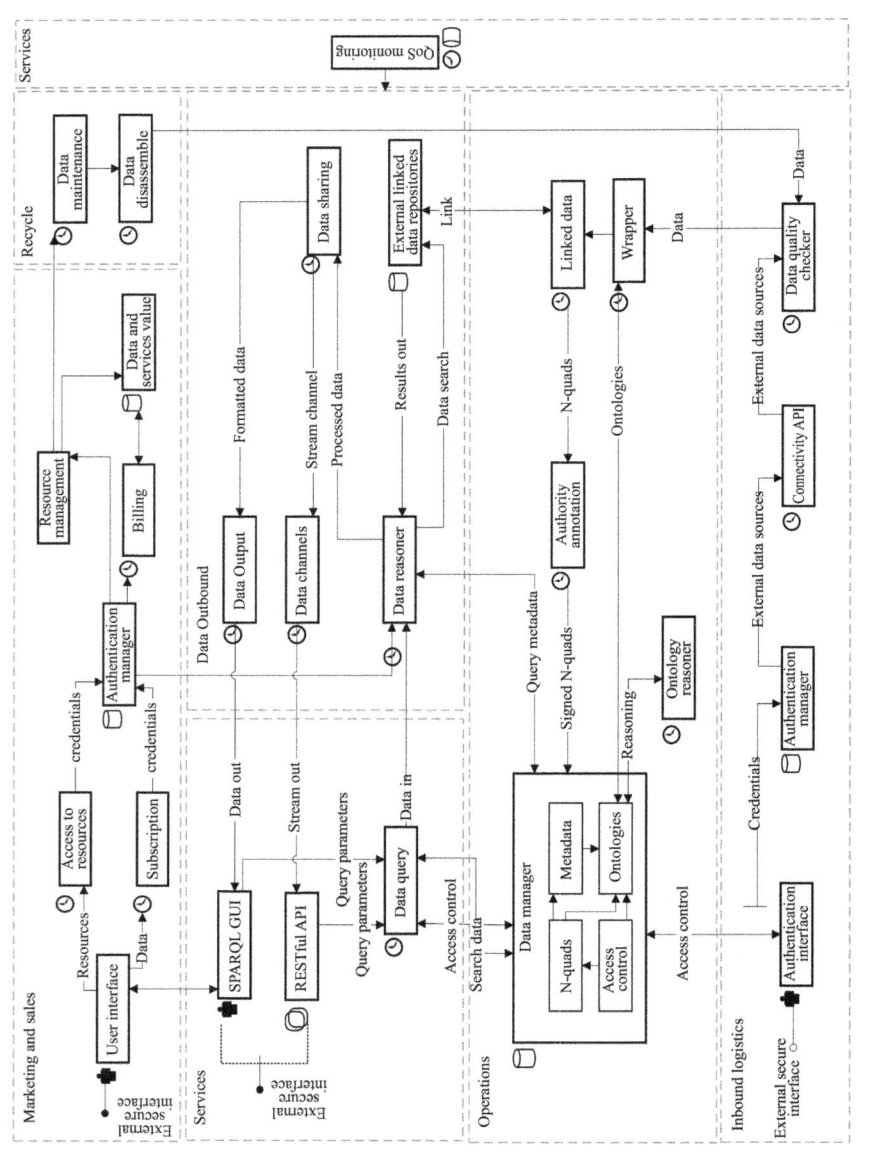

Figure 8.12   Holistic view of the platform and its respective data flows

and request approval to submit city data. They provide valid registration details and wait for registration confirmation.

Platform Providers may authorize or not data publishers to offer both open and proprietary city data in the data infrastructure. In the data infrastructure, data may be manually uploaded or submitted via APIs to support the integration with the Internet of Things. This component provides a confirmation of receipt of data publication to the Producer, which may include a request to resubmit data in the case of errors resulting from the submission. Once data has arrived, it must undergo several reviews, including virus checking, format compliance, metadata minimum requirement agreement, quality and anticipated content and data formatting. The data infrastructure includes the ability to record all actions and decisions made concerning the publication of city data. The reasons for publication failure (e.g. missing metadata information, non-valid dataset) is provided back to the city data publisher. In some cases, the provider can then resubmit corrected data and metadata information, while in other instances data publication refusal criteria should prevent the publisher from submitting the same dataset at a later time period (e.g. in cases of suspicious datasets – copyrights violation, viruses).

When data is successfully submitted (either via APIs or manual upload), it will be processed/prepared for storage into the data infrastructure's database. The data submissions and request for updates are sent to the Data Operations Layer.

- **Functional Requirements:** FREQ.1, FREQ.5, FREQ.7, FREQ.8, FREQ.9, FREQ.10, FREQ.11, FREQ.12, FREQ.17, FREQ.18, FREQ.19, FREQ.20, FREQ.21, FREQ.23, FREQ.26, FREQ.33, FREQ.79
- **Quality Requirements:** NFREQ.14

Data Operations Layer: When data is successfully submitted (either via APIs or manual upload), it is processed/prepared for storage into the data infrastructure's database. This layer is responsible for enriching data with semantic and provenance annotations. The Wrapper component converts multiple formats and databases to well format linked data according to the relevant RDFs and/or vocabularies (e.g. OWL, SNN, SWP). Named graph is an extension of RDF and enables incorporating context/provenance information at both graph and database level [426]. The choice of data serialization impacts the scalability of the platform and will be tested on the architecture simulation. Access to data is granted after identifying users and validate access against eligibility rules (e.g. open data, user has subscribed and has access to consume a piece of data). Two modes of authentication have been selected: user credentials and digital certificates.

User credentials enable users to access authentication interface on the platform, while digital certificates enable humans and machines to publish (POST) and consume (GET) data through HTTP requests. Once data has been properly formatted as TRIG, data is submitted to the External Linked Data Repositories Component which is responsible for linking the data with external linked data sources. The final linked data is submitted to the Authority Annotation Component for further annotation with regard to data ownership and access control policies. The license of a particular data is replicated to any derived data to guarantee data ownership. Data is annotated with access control policies and privacy protection of sensitive information. For instance, data can be annotated as public (data with no access restriction) or private (data with access restriction). The data and its respective metadata, including provenance annotations, are stored by the Data Manager Component which is responsible for data persistence.

We also consider the payload and infrastructure resource demand of data linked with external data sources using and sent to the Authority Annotation component which creates and verifies graph signatures using X.509 certificates. Data is annotated with access control policies and privacy protection of sensitive information. For instance, data can be annotated as public (data with no access restriction) or private (data with access restriction). Access-control levels and license models are associated to data which is subject to restrictions relating to access and conditions of use. The data infrastructure provides database update responses and keeps an audit trail of all actions to enable rollback to avoid update errors to be propagated in the platform.

- **Functional Requirements:** FREQ.2, FREQ.3, FREQ.4, FREQ.5, FREQ.13, FREQ.14, FREQ.15, FREQ.16, FREQ.22, FREQ.24, FREQ.25, FREQ.32, FREQ.34, FREQ.35, FREQ.36, FREQ.37, FREQ.39–46, FREQ.52, FREQ.54–69, FREQ.71, FREQ.72
- **Quality Requirements:** NFREQ.1, NFREQ.3, NFREQ.5, NFREQ.6, NFREQ.7

Data Outbound Layer: The hybrid approach model was the platform strategy chosen. It shares features of a platform which provides real-time data through APIs and bulk download. Data query component is responsible for answer data queries and requests from the SPARQL, which is the standard RDF query language, and from the RESTFul API (via Web Server). This component validates subscriptions and digital certificates (when applicable), analyse data-access control and expose to the user just the information the user has access rights, as well as information about the provenance of the data. The ontology reasoner – enabled by machine learning – allows the inference of logical consequences, derivations from a set of asserted facts in the N-Quad store accordingly with a particular ontology. It is useful to use a reasoner in case one wants to trace where an inferred statement was generated from, infer access-rights propagation in graphs.

Once such verifications are completed, the data reasoner component moves a copy of the data to the data sharing component for further processing. If special processing is required, the data sharing component accesses data in staging storage and applies the requested processes. The types of operations, which may be carried out, include sub-sampling in temporal or spatial dimensions, conversions between different data types or output formats, and other specialized processing (e.g. data visualization).

After this process, the data will be sent to the appropriate delivery channels (e.g. API's, GUI). In the case of private and commercial data, the users are redirected to the data marketplace interface. This layer also encompasses function to verify corruption during any internal data transfer. This function requires that all hardware and software within the data infrastructure provide notification of potential errors and that these errors are routed to standard error logs that are checked by the Platform Provider.

- **Functional Requirements:** FREQ.27, FREQ.28, FREQ.29, FREQ.30, FREQ.31, FREQ.38, FREQ.77

Data Services Layer: Data Reasoner component is responsible for validating data licences, retrieve information about the data consumption and user details, subscriptions and any other necessary validation before exposing the data to the final user. After the data has been validated, the Data Sharing component exposes the data through different data channels (download and streaming channels). Real time data is exposed as streams of data on the API endpoint, and batch data download can be exposed on either the API or SPARQL endpoint. Data services (e.g. data visualization, data manipulation tools) are offered via GUI. Data Facilitators can register in the platform and request approval to deploy services in the platform. They provide valid registration details (to be defined) and wait for registration confirmation.

Platform Providers may authorize or not data facilitators to offer both open and proprietary services in the data infrastructure. Data facilitators are also provided with functions for updating, maintaining and accessing services as well as tracking their usage by users. In case of updates the data infrastructure must log in the database update details and send to service providers a response indicating the status of the update. The QoS component constantly assesses the overall service-level of the platform architecture, including availability, reliability and performance metrics. This component of the data infrastructure will involve the application of artificial intelligent (e.g. fuzzy systems) to guarantee the platform is working under optimal condition. This layer also provides the services and functions for the Platform Providers to manage the overall operation of the data infrastructure. Administration functions include monitoring quality of service agreements, auditing data publication to ensure that they meet archive standards and maintaining configuration management of system hardware and software.

In overall, it provides system engineering functions to monitor and improve platform operations, and to inventory, report on and migrate/update the contents of the data infrastructure's databases. Data facilitators are offered with functions to manage their commercial services. The data infrastructure provides functions for updating, maintaining and accessing both service and its respective commercial transactions. Furthermore, it enables them to track the usage of services by users. Services usage tracking includes performing queries on the platform to generate result sets and producing reports from these result sets.

- **Functional Requirements:** FREQ.47, FREQ.48, FREQ.49, FREQ.50, FREQ.51, FREQ.52, FREQ.70, FREQ.73, FREQ.74, FREQ.75, FREQ.76, FREQ.78, FREQ.81–88, FREQ.98, FREQ.99, FREQ.100, FREQ.101, FREQ.102, FREQ.103, FREQ.104, FREQ.105

- **Quality Requirements:** NFREQ.2, NFREQ.3, NFREQ.4, NFREQ.5, NFREQ.8, NFREQ.9 , NFREQ.10, NFREQ.11, NFREQ.12, NFREQ.13, NFREQ.15–25

Data Marketing Layer: Data and service consumers can register in the platform and request approval to consume city data via GUI or APIs. They provide valid registration details (to be defined) and wait for platform to confirm their registration. Users must accept the terms and conditions of platform usage and define how their personal data can be used by the Platform Owner. Users can manage and alter their registration information at any time they want to. A middleware billing system supports the financial transactions on the platform based on the financial value of the data. Data access is granted after users are authenticated and subscribed to a dataset. User's authentication and payment of data subscription can be made via *smart contracts* using blockchain technologies such as

The financial model adopted in the platform is data-as-a-service in which users are able to subscribe to entire datasets or particular pieces of information in the dataset. The platform must provide an update response indicating the status of the payment. If successful, service is readily available to be used, otherwise the user can re-try the payment or cancel transaction. The consumers of data services are provided with functions which enable them to manage their subscriptions and financial transactions. These functions include updating, maintaining and accessing financial transactions. User's feedback is handled by the Data and Services value components.

- **Functional Requirements:** FREQ.89, FREQ.90, FREQ.91, FREQ.92, FREQ.93, FREQ.94, FREQ.95, FREQ.96, FREQ.97

LOGISTICS: REVERSE

Data and Services Recycle Layer: This is the first silo of the reverse logistics of the data infrastructure. The Data Inbound also provides mechanisms for data publishers to manage their resources. Data publishers are provided with services and functions for updating, maintaining and accessing both data and metadata, licensing models, as well as tracking the usage of resources by users. Data usage tracking includes performing queries on the data management data to generate result sets and producing reports from these result sets.

- **Functional Requirements:** FREQ.80, FREQ.83
- **Quality Requirements:** NFREQ.14

Part IV

Assessment and evolution of data infrastructure design

Chapter 9
The dynamics and evolution of business models

The development process from business idea to established business can be divided into a number of phases, and each phase *'helps to understand the evolution of the competitive landscape in the wake of an innovation or change, and the consequences of such events for firm strategies and business models'* [270,297]. A value proposition should be studied over its entire life cycle [389]. Hence, business models are dynamic rather than static, and it is important to understand how they evolve over time. The role of leadership is to monitor and act on uncertainties which arises as the business models evolve [390] and to ensure the components of the business models are adjusted to address emerging and changing requirements [391]. Cortimiglia's work [392] pointed out that most companies tend to start with designing or improving their value creation components of business models (e.g. value activities and resources) and then innovate all other business model components as the dynamics of the business models are assessed.

The use of a dynamic framework in the design and realization of data infrastructure can help platform owners to understand how their infrastructure is shaped and changed by its environment throughout its life cycle. For instance, Mason and Rohner [393] distinguish four venturing phases in which an organization can develop and offer a new service: Phase I: venture vision (validating the concept), Phase II: alpha offering (building while planning), Phase III: beta offering (testing the concept) and Phase IV: market offering (calibrating and expanding). On the other hand, Afuah and Tucci [297] suggest a three-phase process: (i) the emerging or fluid phase in which the value network is organized, (ii) the growth or transitional phase in which a standard or dominant design is put in place and (iii) the mature or stable phase in which companies focus on keeping and improving their competitive advantages. Based on existing phasing models, Bouwman's research [269] considers three phases in the life cycle of business models for mobile systems [269]: Technology/R&D, Roll out and Market phases. Below, we outline specific tasks that support the management of business models dynamics:

- Monitor the external forces affecting every stage of the business models life cycle.
- Anticipate the potential consequences that affect the critical success factors (CSFs).
- Implement actions designed to modify the business models and the reference architecture so that the CSFs and the smart cities capabilities are either preserved or increased.

In our approach, the dynamic business models are divided into four phases: Research and Development, Procurement, Roll out and Market. We have included the procurement phase as smart cities need to be able to identify and procure the best technical solution for their communities and businesses while demonstrating the local economic benefit of that procurement [394].

Research and development: It consists of the conceptualizations of the service concept and business model. Bouwman *et al.* [269,270] argue that in this phase, R&D (basic and applied research) and technology play a dominant role, the core activities being services and infrastructure definitions.

Procurement: The transition from the first to the second phase is marked by strong engagement with the market to procure sustainable and innovative solutions that will jointly deliver the digital infrastructure strategy. This phase requires intelligent organization (e.g. centralized process management) and professional well-trained staff who should identify what is actually available on the market before deciding whether and what service to contract (e.g. cloud computing). Platform Keystone should work with the DDVN members that will deliver the services or infrastructure so that they can translate their needs, and stakeholders can therefore offer a more tailored service that meets the needs of cities. Companies and the market must understand the city's objectives, strategies and vision and '*must be able to turn them into business cases*' [394]. The procurement must also take into account the engagement of end-users as it is essential to inform citizens about the benefits and their involvement in the overall strategy, hence supporting the human and social capability of cities.

Roll out: After procuring the solutions that will support the delivery of the data infrastructure comes the time to launch it on the market (start of the market phase). Activities in this phase are '*testing service concepts in focus groups, field experiments, the roll out of technology, testing of alpha and beta versions of the service and (small-scale) roll out on the market*' [270].

Market: After experiments prove the data infrastructure strategy is successful, it is disseminated into the market as a tested and robust solution. Core activities in the market phase are '*retaining rather than capturing market share, commercial exploitation on a day-to-day basis, focusing on operations and maintenance*' [270]. The market phase encompasses stages of market offering, maturity and decline, and constant feedback loops occurring so the services provided can be in continuous evaluation and improvement.

9.1   External forces

A business model can be sustained over time by taking into account external factors that influence the model [262]. The nature of external constraints on business models can be technical, economic, cognitive, structural, legal, political and cultural [395]. In the STOF model, market drivers, technology and regulation are external forces with the most direct impact on the business model for mobile services Bouwman *et al.* [270] define. Aligned with previous research, we highlight that for smart cities data infrastructure the feedback of users and niche players also influences the business models throughout its phases. The impact of these external drivers on internal business model components will be different in each phase, and they are discussed in the following.

Research and Development phase: technology will be the major driver behind new business model development. New business models can also be driven by market developments, such as changing customer demand or new entrants on the market.

Procurement phase: technology and regulations will be the major driver behind business models being procured. At this stage, there will be many design choices and service providers to be aligned with the regulations and specifications of the procurement phase.

Implementation/roll-out phase: strict changes in regulations can be put in place by competitors. As a result, it has to be certain that the service complies with such changes in regulation. Feedback of users plays a major role during the roll-out phase. Early adopters can report on the quality of service (QoS) and the usability of the services provided. This feedback can be useful to make the necessary changes to meet the customers' expectations. Furthermore, as technology and services are replaced/changed, it is possible that new niche players join the value network introducing new requirements and requirement balancing activities, especially in the organization, value and governance domains.

Market phase: there should be a focus on retaining customers and users, and updating the business models as necessary to ensure that it is still competitive and the business offering is still attractive to the customer.

9.2   Business models evaluation and refinement

A very important and challenging field of business model research concerns the definition of indicators, business model measurement and evaluation. A number of authors have written on this question providing different approaches to the topic (e.g. [297,306]). Hamel [306] discusses the wealth potential of a business model by analysing four factors:

- The extent to which business concept is an efficient way of delivering customer benefits.
- The extent to which the business concept is unique.
- The degree of fit among the elements of the business concept.
- The extent to which the business concept exploits profit boosters (increasing returns, competitor lock-out, strategic economies and strategic flexibility) that have the potential to generate above-average returns.

Afuah and Tucci [297] assess business models on three levels: (i) profitability measures (earnings and cash flows), (ii) profitability predictor measures (profit

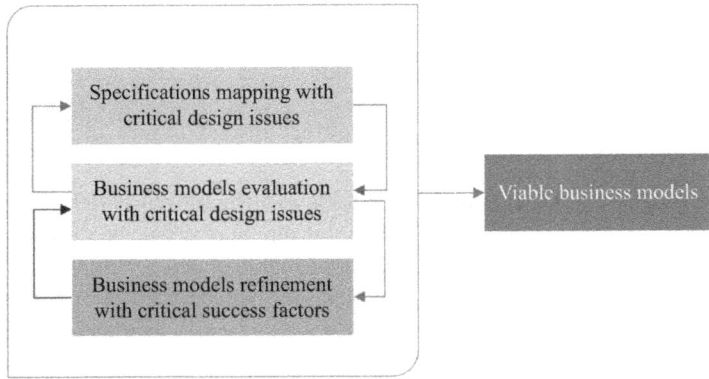

Figure 9.1   The relationship between CSFs and CDIs to create a viable business model (adapted from [270])

margins, revenue market share and revenue growth) and (iii) business model component attribute measures (benchmark questions for each business model components). Bouwman *et al.* [270] argue that understanding critical design issues (CDI) and CSF helps in validating the viability of business models for mobile service. A viable business model should create value for customer and network alike. Value creation for business actors (network value) is complex due to the varying and perhaps conflicting strategic interests of organizations in the network. Although design choices in the technology domain should satisfy the requirements of the service domain, not every solution will be affordable; thus, there are also interdependencies between the technical and financial domains [269,270].

- **CDI** is defined as a design variable that is seen by the practitioner and/or researcher to be crucial to the viability and sustainability of the business model under study. The CDIs found by Bouwman *et al.* [269,270] are based on a number of case studies but may differ according to the case in question.
- **CSFs** refer to '*the limited number of areas in which satisfactory results will ensure that the business model creates value for the customer and for the business network*' [271].

In this research, we define CSFs as the ones that will support the realization of the seven capabilities of smart cities (physical, human, social, institutional, innovation, sustainable and economic). CSFs are based on CDIs and iteratively refined to create a viable business model, as illustrated in Figure 9.1 [270].

9.3   Critical design issues

We have identified potential common and recurrent CDI from the literature that involving the business models of data infrastructures. The descriptive SMARTify model we discussed in the previous section has been used to describe and analyse the business models of a number of cases, including open data platforms, proprietary data platforms and hybrid data platforms discussed in the literature. As there is no such concept fully developed, there is no blueprint for data infrastructure for smart cities, hence our inability to benchmark data infrastructure strategies. However, we argue that materials from the literature and case studies help us to understand what may be the CDIs for each of the five domains of SMARTify.

The CDI defined for each domain was based on materials from the literature and case studies. We will discuss the CDIs in more detail, establishing the CDIs of each domain, and how they relate to balanced business models. Next, the CDIs are used to build causal models describing their relationship with CSFs.

9.3.1 Service design

We define the CDIs of data infrastructure as follows: targeting, value elements, multi-context services, co-design of solutions, reputation and user engagement.

Targeting. An important issue in almost every case was choosing a profitable target group. Should the service offering be targeted at open data or proprietary data consumers? Or both? Should the service offering be targeted at expert data scientists with previous knowledge of data manipulation or should it be also focused on users with no prior knowledge? Should we provide data which complies with the five start principles of open data or just provide it as it is? Sometimes, service providers formulated a growth strategy in which the offered services evolved from one segment to a more advanced segment of data provision. At the time the UK open data store was launched (data.org.uk), the data provided was mainly generated in the formats they existed. As the demand increased for more reusable data, several datasets using reusable data standards (JSON and RFD) have been released.

Creating value elements. Closely connected to choosing a target group is formulating a compelling value proposition for end-users [270]. In the smart cities domain, the added value of a service can be based on value elements such as availability, complexity and feeling of security, trust and privacy. It is the core services plus the support or auxiliary services which creates the customer value. For example, the availability of a mechanism for cleaning up data was found to have a significant influence on customer value as it ensures data is valid, whereas in other cases a mechanism to integrate and augment data was of more importance [132,230,369,396]. Quite a number of the studied smart cities platforms did not have a clear and compelling value proposition and seemed to be blinded by the technical possibilities, that is, instead of seeing technology as an enabler they have taken it as the most important part of the solution (e.g. [96,181]). The cases indicate that there is a clear tension between the possibilities offered by technology and the expectations and needs expressed by end-users. Several works have reported the need to provide services in which any kind of user can engage and participate. Complexity also represents a value element. This sense of easy to use develops the human capital of cities in which users are more inclined to engage and participate in civic initiatives.

In most smart cities platforms, an important value element was trust, feeling of security and privacy. In some cases, trust had to do with the reliability of data and data providers business actors, whereas in other cases, trust was associated with the security and privacy of the technology that was deployed (see [192]). Different measurements have been suggested in the literature to enhance trust, security and privacy preserving issues in smart cities platform, as discussed in the technology domain. In business a certain level of trust between agents is indispensable so that business can take place. *'Trust of a party A in a party B for a service X is the measurable belief of A in that B – a proposition in a design science approach will behave dependably for a specified period within a specified context'* [397]. This shows that the notion of expectation is central to the concept of trust, which has been based on identity, assumed quality or the perception of risk and it deepens over the time of a relationship [398]. But in a business environment that has become increasingly global, transactions more and more virtual and where the implicated parties do not necessarily know each other anymore before conducting business, new trust mechanism have gained importance. Modern digital technologies offer a large range of innovative or improved mechanisms to build trust in e-business environments [399] by improving the expected output of a transaction. For instance, virtual communities are a powerful but two-edged instrument of trust.

Besides content officially published by a company, the members of a virtual community of transaction compare and aggregate their experiences and thus give a perspective independent of vendors and advertisers [359,400]. However, companies have a very limited direct influence on virtual communities and often fear their power. For data infrastructure, we expect that trust will be built upon the provision of services and data that will be perceived as trustable given the feedback provided by the users.

Multi-context services. It refers to the provision of data for the different contexts of data usage. As explained in the expected value of the service domain, there are five distinct processes of city data use [320] which are not mutually exclusive and many users of government data (open) employ multiple usage patterns. By providing multi-context services, data infrastructures will serve a larger number of stakeholders of city data and will increase the value return for releasing data (see value domain).

Co-design of solutions. Platform owners have to make decisions about establishing competitive or collaborative provision of services. For instance, by enabling developers to link their applications to the platform (e.g. visualization tools), allowing data tool providers to offer data cleaning or mashup services. Enabling solutions to be developed by multiple stakeholders can lead to increased value of services to the customer of data infrastructures. The co-design of solutions is determined by the Platform Openness strategy.

Reputation. Reputation was found to be an important issue in relation to reaching the customers who had been targeted. Reputation seems to have a direct influence on the perceived value of service offerings, which makes it an important means to create customer value. Data infrastructure providers may decide to promote a new service by bundling it with an existing product, which carries either the same brand name or a different brand name. For instance, using Google Refine to assess the quality of data enhances the trustworthiness of the platform as the tool for quality verification belongs to a well-known organization. An important requirement for reputation is recognition by the target group and the existence of a match with the intended value proposition. Reputation constitutes a pivotal resource for generating and sustaining competitive advantage as they strengthen relationships and help creating a distinction among entities that may satisfy a costumer's need [401,402].

User engagement. In addition to choosing a target group and defining the added value, user engagement was found to be a CDI. Customer retention refers to marketing strategies aimed to keep customers satisfied and loyal with the product or service. The cases show that service providers adopt different strategies to stimulate recurrent usage of their services. Personalization, accuracy and actuality of data have been acknowledged as cases to attract and retain users [192].

9.3.2 Technology design

Based on our case studies, CDIs that originate from the technology domain are data/metadata, security, QoS, usability, content usability, content reusability, interoperability and accessibility.

Data/metadata features. The new world of multi-structured data has imposed many challenges to data processing mechanisms, being too difficult, too time consuming or too expensive to analyse. The architectural features of city data must be taken into consideration while designing data infrastructures. As discussed in the technology domain, different measurements should be considered and employed to enable a data infrastructure strategy in which data complies with the architecture features of city data.

Usability. Companies must also think of new and innovative ways of making the life of users as simple as possible. Reducing users effort to consume city data means creating value through lower search, removing a number of poor data which require intensive data manipulation and manual processing and verification. One of the most recognized impacts of digital technologies on value creation has been on the reduction of customer efforts.

Numerous city data portals – including the ones commented in this book – have trusted on the Internet as a channel for convenient sharing of information, but not all of them have fully understood the consequences. Although they have provided data through web portals in an attempt to reduce the effort of users in obtaining data from different sectors in one single portal, most of them have taken their current infrastructure as a starting point, and neglected other business model elements. The user needs for finding, processing and using data have not been taken into consideration, there is no widely accepted standard for expressing the syntax and semantics of the city data [199], and current release of data which requires

Table 9.1   Berners-Lee's open data scoring

Scoring	Rationale
★	Make your stuff available of the Web (whatever format) under an open license
★★	Make it available as structured data (e.g. Excel instead of image scan of a table)
★★★	Use non-proprietary formats (e.g. CSV instead of Excel)
★★★★	Use URLs to identify things so that people can point at your stuff
★★★★★	Link your data to other data to provide content

substantial human workload to prepare the data for machine processing and to make them comprehensible [192,196–198]. However, if the data infrastructure channels are soundly integrated into a business model they can have remarkable impacts as the key elements will be emphasized.

Content usability. Refers to how usable a particular dataset is by users and machines. There are different ways in which content usability can be measured, and they are, to some extent, subjective. One of the more common methods is the open data five-star scoring mechanism suggested by Sir Tim Berners-Lee (http://5stardata.info/), which is used by data.gov.uk is shown in Table 9.1.

The Data.gov.uk portal has recently provided information on the proportion of datasets attaining each star rating (beta form). These figures indicate that, in the last year, nearly a quarter of datasets have attained a three-star rating. Less than one per cent of datasets have been given a five-star rating, indicating they are linked data (RDF format). Over 50% of datasets have attained no stars – though this is due to many datasets not yet being rated.

Content reusability. There is scope for improvement, especially around the greater provision of data and metadata including quality, format and licensing. Content reusability concerns to the degree in which data is made available and is re-usable. Although open data has no restriction to its usage, there are different license schemes for open data [349]. In the case of private data, the restrictions of usage and distribution are not general but are decided by the individual businesses. Intermediaries play a crucial role in the use and re-use of data – converting data into information that becomes a catalyst for action. Data infrastructure must make available for the re-users of city data all the information about data ownership and under which circumstances data can be re-used, augmented, integrated, republished or commercialized.

Security. Trust of end-users and customers in a service offering is partly determined by the way security is implemented in the technical architecture. That is, the way in which access to a service is granted and how security of communication and data is realized. Security may require a trade-off between ease of use or privacy considerations and preventing abuse. While anonymous access guarantees privacy of users, traceability of users due to abuse of the service is not possible. Authentication of users or blockchain-based interactions may solve such problems but should be examined if all users should be authenticated or just the ones consuming data with license restrictions or subjected to a fee.

Quality of service. In all the cases we studied, the performance of the technical architecture in delivering the technical functionalities is a major concern. QoS seeks for providing an intelligent environment of self-management components based on domain knowledge in which the components of the data infrastructure can be optimized easing the transition to an optimized service level. Because the satisfaction of requirements in one QoS feature often requires certain sacrifice in other QoS features, trade-offs of requirements among multiple QoS features must be taken into account in the design data infrastructures.

The emergence of the cloud computing paradigm brings a new set of QoS open issues: accomplish the Service Level Agreements (SLAs), predict future workload, process large and diverse data streams and transactions, make real-time decisions, reasoning and inference, dynamic adaptation and provision of resources and services, among others. New semantic technologies have emerged as an option to design and develop intelligent software components to fulfil applications requirements (see [403]).

Interoperability. The adoption of the service is in part determined by the extent to which (i) services can be integrated into the existing platform structure and (ii) data is interoperable. City data interoperability is a fundamental requirement for the design and realization of smart cities. Nevertheless, the high volume, different velocities in which the data arrives, is stored, and retrieved, the disparate heterogeneous sources which originate a variety of different data formats, the variability in the meaning and context, the data vulnerability in terms of security, trust and privacy, the lack in verification of the data which may present poor quality and redundant affecting the veracity of the data, the issues surrounding validity and legal conditions of the data usage and re-use, and the difficult task of integrating the data to unlock value has posed a significant challenge in the integration of events that will trigger city services. Platforms which solve services and data interoperability issues will foster service innovation by enabling services providers to develop new services reusing existing functions.

Accessibility and Fairness. The accessibility of the service to the target group is obviously influenced by the choice of platforms, architecture and services. The distribution channels which will make the data available for use are the connection between the data infrastructure and the targeted customers. A distribution channel allows a data infrastructure to deliver value to its customers, either directly, for example, through a providing data via application program interface (API) or bulk downloads, or indirectly through intermediaries, such as data integrators or platform data services. It demands a strategic perspective that views channel decisions as choices from a continually changing array of alternatives for achieving market converge and competitive advantage. Its purpose is to make the right data and/or services available at the right time to the right people – subject of course, to the constraints of cost (e.g. developing APIs), investment and flexibility (e.g. hybrid platform model). Accessibility of the service via data distribution channels was found as a CDI for data infrastructures. A hybrid approach model to the delivery of city data seems to be appropriate. It also involves trade-off between flexibility and costs. The costs for building on legacy systems may be lower but provides for less flexibility than an open system based on standards and open interfaces. As explained in the governance domain (platform strategy), it supports the provision information on demand as it shares features of a platform which provides real-time data through API's and bulk download and therefore can fulfil the requirements of data consumers who want to consume historical data and the data facilitators who want to create real-time applications.

9.3.3 Value network

Based on our case studies, CDIs that originate from the value network domain are broaden partnership, network complexity and partner selection.

Broaden partnership. Partnering between competitors in order to create new markets or to achieve common standards is not uncommon today. Many of today's technologies are based on network externalities and the winner often takes all it is advantageous to form consortia of partners [270]. The potent mix of specialized expertise with the disciplining power of markets can foster innovation at a rate that exceeds by orders of magnitude conventional business models. As discussed in Chapter 2, products that became platforms from 1990 until 2004 enjoyed a 500% increase in innovation, most of which came from outside developers [366]. A platform's success therefore depends not only on the platform owner but also on a multitude of ecosystem partners' ability to deliver [404].

Network complexity. Network complexity may arise from the number of relationships a focal business actor needs to manage in a value network and from the effort needed to connect stakeholder's services in the technical architecture. As there is no data infrastructure which enables partnerships in the provision of data, our understanding on the network complexity in such environment is limited. However, according to studies on mobile services [270], it is necessary to trade-off between the need to reduce complexity in the network (number of stakeholders providing services) and the need to have access to critical resources and capabilities (e.g. customers, content and funds, among others).

Partner selection. An important design issue in all cases is acquiring access to resources and capabilities needed to realize a service offering. Firms need to decide whether to outsource certain activities or to perform them in-house. A distinction can be drawn between business actors that provide indispensable and irreplaceable (critical) resources and capabilities, and those who provide supporting resources and capabilities. The platform model essentially outsources to thousands of outside partners innovation that used to be done in-house, who take responsibility for risks and innovation and then share their innovative application with the platform owner. Network complexity and access to resources and capability have seen to be crucial when selecting partners to extend the services of a platform [270].

9.3.4   *Value design*

Based on our case studies, CDIs that originate from the value network domain are demonstration of impacts, pricing, division of investments and risks and division of costs and revenue.

Demonstration of impacts. Value can be generated through its independent use or re-use or by combining city data with other data sources to become important inputs into products, services, decision-making and other outcomes. As discussed previously in this chapter, city data acts as an input as a data point for complex algorithms and analyses, as a source of insights (e.g. dashboards or visualizations), as API feeds for apps and so forth. Individuals with the technical skills and interest to work with data may decide to use their skills to derive insights from data (research or other purposes) or present it in a form that is more accessible to non-technical users. In these cases, the monetary value generated by the use of public sector information may be small or non-existent, but the product may produce wider social, sustainable and economic benefits, as discussed in the literature review.

The use and re-use of city data can generate a range of benefits, which in turn can create value for consumers and producers alike. Some of the types of benefits include the value of reduced carbon emissions, reduced fuel use and time saved due to reduced congestion through using apps (e.g. live transport updates), cost savings arising from analysis of different public sector information datasets, improvements to decision making and choice due to new insights from city data which helping generate economic growth, and better policy making using city datasets that improve value for money and the efficacy of policy [161,192,344,345].

Pricing. With regard to the adoption and actual use of a service, the perceived customer value must at least equal, and preferably exceed, the price of a service. In essence, opening up data is not always free, and there are some potential costs associated with the production and presentation of open data that need to be considered and accounted for. Such matters raise the need to take a special care when designing platforms for city data management. Such platforms must be designed in reasonable and affordable ways, which do not add unnecessary financial costs and potential loss of revenues. There is a substantial commitment and investment on the part of public agencies as they need to acquire new skills, train employees, get data specialists to prepare information to be released, purchase technologies and upgrade network infrastructure, which need to be accounted for.

An alternative to recover costs of opening up data is to charge user to consume data and services, or allowing free non-commercial re-use and charging for commercial use and re-use on the basis of a license. Converting public data can have cost implications, particularly if there is a high-level use of proprietary and legacy systems. Initiatives such as converting government data to semantic web and linked data formats can be time consuming and therefore costly, similar to protecting partial pieces of data on datasets. These additional costs can lead to some reserve on the part of data publishers to open data. With regard to the charge of services provided into the platform, users should be able to define the price/tariff of the services and data they are providing. Charging users to consume data may increase competitiveness on the platform, in which data providers will compete in providing the best quality data to end-users. Data consumers may opt to consume data that is private and there is a cost associated to it or strictly to consume open data that is free of charge. The platform must support both financial models. The issue of charging for city datasets and their funding models remains an open and complex issue.

Division of investments and risks. There are financial risks involved in developing and introducing a new service, as there is uncertainty about the resulting return on the investment. An alternative solution is to invest on a long-term solution and design databases with the right of public access in mind (e.g. one-way encryption on databases). This would enable data providers to release data at low costs, and as a consequence, increase its consumption by data consumers/data enablers/knowledge creators.

The revenue of opening up private and public data goes beyond financial terms, yet it involves the reduction in the cost of data re-use and innovation by businesses and individuals. Therefore, part of the cost involved in processing and analysing city data can be outsourced to interested entities. Apps, innovative services and crowd-sourcing provide a cheap way of processing city data and deriving insights that can be used to improve the physical infrastructure of cities. Part of the investment in human resources such as data scientists can be transformed into investment for modern digital technologies that could manage and operate the physical infrastructure. The issues and difficulties involved in releasing open data can be mitigated by, for instance, providing services that enable users to convert their data into linked data format. There are several open-source applications that enable such tasks, for instance, D2RQ which converts relational databases into the linked data model.

Division of costs and revenue. We found that the division of costs and revenues was different from case to case and that in each case it may follow a different logic, e.g. cost based or value based. The relationship between costs and revenues for each stakeholder involved seems to depend on their individual access to critical resources, the services they provide, their expertise and benefits and the risks and level of investments.

9.3.5   Governance design

Based on our case studies, CDIs that originate from the governance domain are appropriate structure, platform openness, follow up on civic engagement and follow up on services provision.

Appropriate structure. Partnerships are voluntarily initiated cooperative arrangements between two or more independent stakeholders to carry out an activity jointly and they are based on a commonly negotiated terms and conditions. The goal behind many partner agreements is the optimization of an organization's operations and services. This can take the form of outsourcing but also shared infrastructure [405]. By entering these agreements an organization can directly benefit from its partner's or supplier's economies of scale or of its specialized knowledge, which it could not achieve on its own. Decision makers and providers of data infrastructures should reflect on what kind of partner resources could leverage their business model and their own competencies.

In most smart cities platforms, we found a single stakeholder (dominant) managing the value network. Current solutions are centrally controlled and exclude the public outside the organization boundaries as part of the data processing system – either by offering data or just providing feedback. It is very likely that platform Keystones will still set the rules with regard to collaboration (organizational arrangements), and monitored compliance with these rules. An appropriate structure adopted by the Keystone may enable complementors to co-operate and co-exist into this complex ecosystem, as the ecosystem (stakeholders) mechanisms and regulations may be orchestrated. Platforms and the ecosystems (people, technology and organizations) that grow up around them have the power to *drive innovation and transform industries* [61]. They form the context for both cooperation and competition, in ways that potentially combine the best of both [307].

Platform openness. The level of openness indicates the degree to which new business actors can join the value network and are allowed to provide services to customers. As the complexity of service grows, it is increasingly difficult for a data infrastructure to specialize simultaneously in all domains that go into producing valuable and usable data. In organizations, there has been a disaggregation of firms into complex supply chain networks involving many partners, and now into even larger ecosystems of smaller firms that specialize deeply on their products and services. These outsiders can potentially bring insights about specialized domains, different application markets and geographies that one company would struggle to maintain in-house [207,217].

A data infrastructure-centric approach enables pooling of multiple firms' knowledge bases that are more valuable in combination than in isolation. Platform-centric approaches have been enabled by software and services digitization, embedding of processes in software, the emergence of the Internet of Things and ubiquity. No data infrastructures were found of an open model in which partners are free to join the value network and offer services and content. When choosing between various degrees of network openness the desired control, exclusiveness and customer reach of the service were found to be of main strategic concern. The higher the desired level of control and exclusiveness is, the more likely the partners are to adopt a closed model. On the other hand, reaching many customers may be an argument in favour of choosing an open model.

Follow up on civic engagement. There are gaps in the supply of data to both people with expertise and with limited knowledge in data manipulation and processing. Thus, it makes necessary to lower the knowledge required for data access, i.e. provide means to people to easily discover and share data, in order to achieve a large-scale dissemination. Hey and Trefethen [193] point out that an e-infrastructure may facilitate faster, better, different scientific research competences and data usage. Such an infrastructure should provide means for data discovery, assessment, provenance, analyse and sharing.

In order to enhance the data access, integration and dissemination in an urban environment, cities' stakeholders should be provided with a platform to easily discover all types of city data. The general scarcity and increasing competition for skilled workers, especially data scientist, can make it harder to construct the infrastructure for world-class city datasets and scale-up efforts to exploit its value. It can be due to the cost (financial and human effort) involved in understanding how to decrease the complexity of services and lower the knowledge required for data access and manipulation. However, there are some low-cost solutions that can be explored to quickly and cheaply improve the skills base to be able to effectively manipulate and extract value from public sector information, such as promoting the use of free online learning platforms. The success of a data infrastructure will be co-determined by the value perceived by users and consumers. The opening of systems provides the opportunity for creating feedback loops in which the government can learn from the public. New governance mechanisms, capabilities and processes are necessary for dealing with these feedback loops. The nature of the response depends on the available organizational arrangements that make a response possible [210].

Follow up on services provision. There are significant data gaps when it comes to open and proprietary city data. This lack of data can, in some cases, lead to inertia with certain datasets not be released or conversely, undue attention being given to datasets that are unlikely to generate significant value but have a low cost of dissemination. There are a number of routes to addressing these data gaps. These range from a detailed, regular audit of city datasets to improve tracking of current usage. It further reinforces the importance of the feedback loop the platform must support in order to enable service providers (e.g. data/knowledge providers) to understand the value customers perceive from the product they are delivering, and how they can change their strategies so customers can be satisfied.

9.4   Critical success factors

The viability of a business model is determined by the creation of value for the customers and for the organizations in the business network [270]. They define CSFs as *the limited number of areas in which satisfactory results will ensure that the business model creates value for the customer and for the business network.*

In our work we define the CSFs of data infrastructures as the ones that support the seven capabilities of cities discussed in Part I of this book. Wallin [406] describes capabilities as repeatable patterns of action in the use of assets to create, produce and/or offer products and services to the market. Thus, a firm has to dispose of a set of capabilities in order to provide its value proposition. The CDIs are instrumental in the value-creating process as they serve as the starting point for the CSFs causal models. Figure 9.2 illustrates the business models assessment based on CDIs and CSFs.

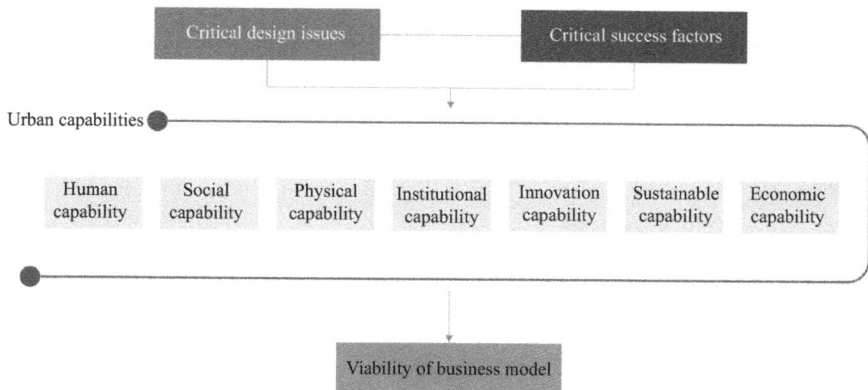

Figure 9.2   *Business models assessment based on CDIs and CSFs*

Table 9.2   Urban capabilities as CSFs of data infrastructures

Capabilities	Rationale
Human	Intelligent individuals, innovators, creative and decision makers who have enormous impact on determining whether cities thrive or wither.
Social	Creative class made by talented people higher skilled individuals that makes communities to evolve towards higher order harmony, collaboration and innovation
Physical	City's endowment of hard infrastructure which are perceived as sources of value creation and competitive articulation in urban economies
Innovation	Application of better solutions that meet new requirements or existing market needs, which is dependent upon the social, human, institutional and physical capabilities of cities
Institutional	Mechanism and structure of a city that can help support businesses and citizens in their quest for optimum intellectual performance, life, mobility and innovation, and the overall city performance can thereupon be achieved
Sustainable	Sustainable, efficient and liveable models of urban development that preserves and originates greener sources of natural resources
Economic	Creation of resources from cost-effective government transparency projects, improvement of services, creation of new and innovative services, stimulation of competitiveness and knowledge developments including a more active participation and empowerment of citizens, and the re-use of information across multiple processes, systems, value chains and stakeholders

In Table 9.2, we summarize the definitions of each urban capability which will be used to formulate design propositions that describe the relationships between the CDIs and CSFs. We have created a break-down structure that explains how business model viability can be influenced by more concrete, design-oriented variables that influence the success factors regarding creating urban capabilities.

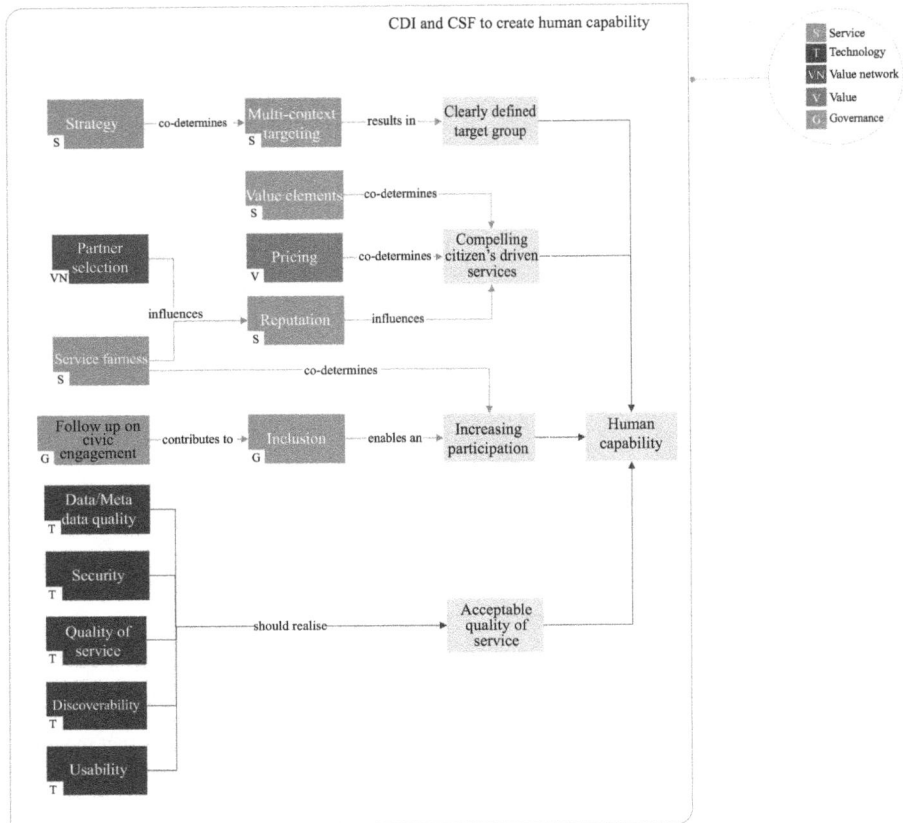

Figure 9.3   Creating human capability CDIs and CSFs

9.4.1   Providing support to human capability

A number of CSFs with regard to viable data infrastructure business models exist as requirements for creating human capability in smart cities. We assumed that high scores on these success factors will result in a service that supports the human capability of smart cities. Figure 9.3 illustrates the CSFs for creating human capability. In this figure, the grey boxes represent the CSFs, while the boxes shown in colour refer to the CDIs of the five domains of SMARTify (see the legend at the top right corner of the figure).

A **Clear Defined Target Group** enables the service provider to stay focused on the user and it is based on the value proposition (service domain). For data infrastructures, it requires focusing on what creates value from the point of view of the customer, the multi-context concept (service domain), instead of the possibilities of the technology. The CDI creating value elements (service domain), reputation (service domain) and pricing (finance domain) enables the creation of a second CSF **Compelling Citizen's Driven Services**. Reputation can be used to differentiate the value proportion from those of competitors [407]. A Clear Defined Target Group and a Compelling Citizen's Driven Services are interrelated in a way that by targeting and specializing in providing services to a market segment a better product (compelling services) can be created and delivered to the user.

Partner selection (value network design) contributes to the reputation of the data infrastructure, hence the importance of having clear and efficient platform governance so that suitable partners – the complementers – of the platform comply with the quality expectation of the platform owner. Following up on civic engagement (governance domain) enables both platform and services providers to obtain feedback on the quality and suitability of the services provided through the data infrastructure. This in turn provides support to users to be part of the data decisions and decision makers can learn from the feedback provided by users of the data infrastructure. Through the adoption of an efficient mechanism to follow up on civic engagement and on the value perceived by users enables the creation of a data infrastructure that can be continuously improved and as a consequence there is an **Increasing Participation** in content use and re-use, and take up from users [165,192]. Studies have suggested that users have become independent entities with regard to the creation and distribution of products and media, which also suggests a power shift from traditional industries towards the people [408], and that innovation is no longer uniquely developed top-down, but are increasingly shaped and moulded bottom-up [409].

While the previous CSFs relate to the service, finance, governance and organization domain, the third CSF is related to the technology domain. An acceptable QoS delivery (technology domain), data and metadata quality, security, discoverability and usability are required because, as far as services are concerned, the quality of the services, the infrastructure, the security in services and data communication and provision, and the discoverability of contents and ease of use aspect of the data infrastructure is as important as that of the service outcome [410]. Data services will be delivered via technology; hence, the CDIs that are related to the technology domain should lead to an acceptable quality level. The QoS also relates to the performance of the technological architecture in delivering the functionality. The security deals with the access to the service and the security of communication and information processing.

The CSFs to provide support for creating human capability are outlined below. It is assumed that high scores on these success factors will result in a service that meets the user expectations and will support the intelligence, innovativeness and creativeness of humans. A service that provides support for creating human capability in the long run can be expected to result in a viable business model

- Clear Defined Target Group
- Compelling Citizen's Driven Services
- Increased Citizen's Participation
- Acceptable QoS

9.4.2 *Providing support to social capability*

A number of CSFs with regard to viable data infrastructure business models exist as requirements for creating social capability in smart cities. Figure 9.4 illustrates the CSFs for creating social capability. In this figure the grey boxes represent the CSFs, while the boxes shown in colour refer to the five domains of the proposed business models (see the legend at the top right corner of the figure).

The first CSF is **Clear Defined Target Group** which is enabled by the provision of multi-context services (service domain) and data which will help the data infrastructure to meet great part of the demand for city data. The demand for city data is likely to increase as users are informed with the demonstration of impacts (value domain) that such data and services may have in their community. The use and re-use of city data can generate a range of benefits, ranging from reduced carbon emissions, reduced fuel use and improved mobility [161,192,344,345]. Such benefits have the potential to create the second CSF **Increasing Sense of Community** in which users are more aware about what is happening around them, such as the energy consumption in their neighbourhood, how well urban services have been provided to their community, among others. This implies that although ICT is a core enabler of smart cities, user-oriented applications and service delivery may further strengthen community development potential [27,411,412]. The success factor increasing sense of community is supported by appropriate structure (governance domain) which facilitates the provision of the right data and platform services the users need and as a result there will be social inclusion (governance domain) for the humans in the city.

The concepts of social co-operation (governance domain) and people-driven innovation can be enabled by introducing the principles of openness, realism and empowerment of users in the development of new solutions [110]. These are fundamental drivers of innovation and problem-solving capability, outlining the intelligence of cities as a collective rather than an individual achievement [111]. These elements create the third CSF **Optimized Collective Intelligence** which is developed as a result of the embedded routines of social co-operation allowing knowledge and know-how to be acquired and adapted. Research has demonstrated that city co-designing of services with the citizens can lead to more sustainable and effective e-services, providing a higher level of citizens' satisfaction and thus higher rates of user take up. Increased accessibility (technology domain) alongside value elements (service domain), such as trust and privacy preserving, creates a cultural change in the sense of using and exploit the wisdom of the crowds: recruit citizens as human sensors through crowd-sourcing (technology domain) and make use of user-generated content [375]. It enables the creation of a fourth CSF **Unobtrusive Participation in Urban Services**.

To conclude, social capability developed in smart cities includes social inclusion of various urban residents in public services, crowd-sourcing initiatives, and educational institutions and R&D capacities. The CSFs to provide support for creating social capability are outlined below. It is assumed that high scores on these success factors will result in a service that meets the user expectations and will enable

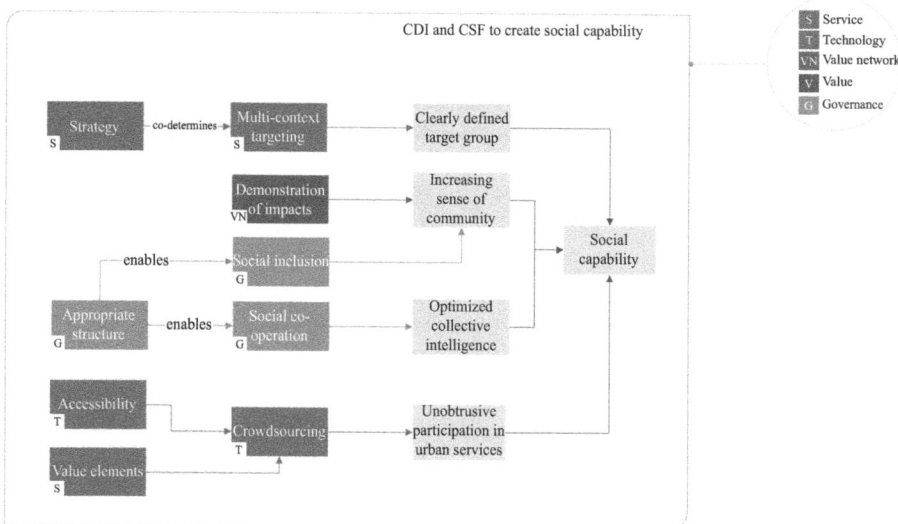

Figure 9.4   Creating social capability CDIs and CSFs

cooperation in knowledge and innovation. Social capability is a resource whose supply increases rather than decreases through the use and, unlike the physical capability, does not becomes depleted if not used. A service that provides support for creating social capability in the long run can be expected to result in a viable business model.

- Clear Defined Target Group
- Increasing Sense of Community
- Optimized Collective Intelligence
- Unobtrusive Participation in Urban Services

9.4.3   Providing support to innovation capability

Innovation means that a firm introduces either a completely new product or service or a revolutionary combination of products and services. Recent research has shown that consumers highly valuate innovation and would be willing to pay for new value propositions [413]. One of the keys to innovation is distinctiveness and impact, which often implies changing the rules of the game and bringing new players into the fold who were not initially considered to be part of the game [414]. Appropriate structures (governance domain) enable the engagement of partners willing to invest and provide the financial support (value domain) that will drive the creation of the first CSF **Local Experimentation and Innovation**. This CSF can be created as a result of the development of social and human capabilities, alongside smart spaces and technologies, which amplifies local creativities, networking, experimentation and innovation. The city gains innovation capability, which is translated into increased competitiveness, a better environment, more jobs and wealth [112].

The demonstration of impact (governance domain) which drives the co-design of solutions (service domain) facilitates the creation of the second CSF that is the **Open Innovation**. Open Innovation scholars suggest the need for smart cities initiatives to transcend their boundaries by sourcing knowledge and technology externally. The underlying mechanisms for accessing external knowledge and fostering open innovation encompasses a range of alternatives including alliances, licensing, open platforms and participation in various development communities [112,415]. These are standards which have emerged to meet the needs of communities and businesses [54]. Open Innovation systems may be especially valuable in promoting a high quality of social interactions (e.g. within communities) [229] as such interactions may involve citizens in public life or collective decision-making or may strengthen a city's participation level in civic engagement.

The city local system of innovation is enhanced by data infrastructure, interactive tools and embedded systems that is provided by a broaden partnership (organization domain) enabled by the openness of the digital infrastructure platform (governance domain). The city gains innovation capability which facilitates **Innovation Transfer**, the third CSF. User engagement (organization domain) has increased the demand for multi-context services (service domain) that give users data and services that facilitate the emergence of smart clusters and innovation ecosystems of entrepreneurs and people that leverage the third CSF **People-driven Innovation**. [416–418]. Multi-context services for innovation creation are supported by interoperability, content usability and content reusability (technology domain).

The competitive advantage in smart cities is co-originated from innovative value-added services on top of data and providing opportunities for innovation and the creation of new businesses through integrated data. Follow up on civic engagement and on creativity (governance domain) enables data infrastructure providers to understand which kinds of innovations are being created and what kind of data and services it is demanding. The gathering and processing of feedback information enables the creation of the fifth CSF that is **Continuous Innovation** [415].

The CSFs to provide support for creating innovation capability are outlined below. It is assumed that high scores on these success factors will result in a service that meets the user expectations and will create a framework for open [415], continuous [419] and systemic [420] innovation. A service that provides support for creating innovation capability in the long run can be expected to result in a viable business model. Figure 9.5 illustrates the CSFs for creating innovation capability.

- Local Experimentation and Innovation
- Open Innovation
- Innovation Transfer
- People-driven Innovation
- Continuous Innovation

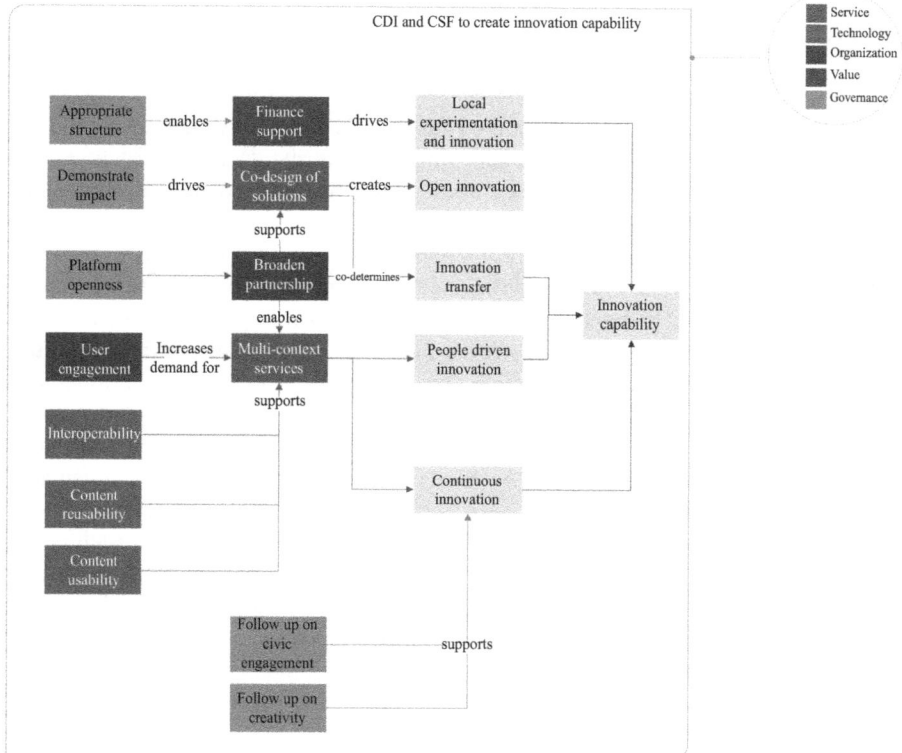

Figure 9.5   Creating innovation capability CDIs and CSFs

9.4.4   Providing support to institutional capability

Economic historians suggest the economic growth until mid-eighteenth century was mainly due to institutional change [33]. Although there was no change in technology at that time, economies managed to grow while at the same time maintained peace, law and order, improved communications and trust, among others [34]. The economic growth and the shape of cities was determined based on trade and on the improved distribution of labour and land, which was enabled by institutional changes [35–37].

Institutional capability deals with the mechanism and structure of a city that can help support businesses and citizens in their quest for optimum intellectual performance, life, mobility and innovation, and the overall city performance can thereupon be achieved. We suggest that institutional capability enable cities to form, develop and use innovation capability and social capability in a coordinated way [131,421]. It is based on governance and policy, because the cooperation between stakeholders and institutional governments is very important to develop and to intervene efficiently to meet challenges posed by a globalized economy and society. Active cooperation has the power to shift the emphasis from the power to decide to the power to transform (i.e. deliver), key to overcoming the usual delivery deficit [48].

We argue that an appropriate structure (governance domain) enables a fair division of investments among the stakeholders (e.g. to bear the costs of data processing services, cloud computing) results in the first CSF **Acceptable Risks**, which in turn will be of benefit to all the stakeholders who invest in the data infrastructure (e.g. governments, research councils). Appropriate structure also regulates the division of costs (value domain) and the finance support (value domain) to enable the second CSF **Acceptable Profitability**. The Platform Openness (governance domain) enables the co-design of solutions (service domain), which is required to deliver multi-context services (service domain) demanded by and increased accessibility (technology domain) of users to the data infrastructure. However, the co-design of solutions influences the network complexity as new participants join the network to provide additional services to the platform. The platform governance (governance domain) strategies have the power to orchestrate the network and enable the niche players of the platform to co-exist and cooperate. It has been currently done in some cities, for instance, The Future Cities Catapult in the UK are working towards bring the stakeholders of smart cities together to understand the requirements and expectations of each party, and the Open Data Institute is trying to understand how cities could better deliver their open data strategies. The British Standards Institution (BSI) has started working as regulator in order to ensure common standards and regulations are being nationally adopted and properly used. Such initiatives which aim at engaging stakeholders in decision making and public/social services enable the creation of the third CSF **Clearly Defined Strategies**.

The cities of the future will be very dependent upon technological innovation and people's creativity, and strong institutional leadership and organizational capacity can provide the best possible conditions to increase cooperation in knowledge creation (social capability), innovation and competitiveness as well as sustainability. We suggest that follow up on services provision (organization domain) and follow up on civic engagement (governance domain) will enable data infrastructure providers to assess how urban services are provided in the cities, and whether the data and services provided through the data infrastructure are meeting the demand from the users who will jointly create the applications and innovations that will support the cities of the future. It enables the creation of the third CSF **Continuous Engagement with Stakeholders** and the fourth CSF **Cooperative Knowledge Creation and Innovation**.

The CSFs to provide support for creating institutional capability are outlined below. High scores on these success factors will create mechanisms that regulate knowledge flows and co-operation in learning, innovation and service provision. A service that provides support for creating institutional capability in the long run can be expected to result in a viable business model. Figure 9.6 illustrates the CSFs for creating institutional capability.

- Acceptable Risks
- Acceptable Profitability
- Clearly Defined Strategies

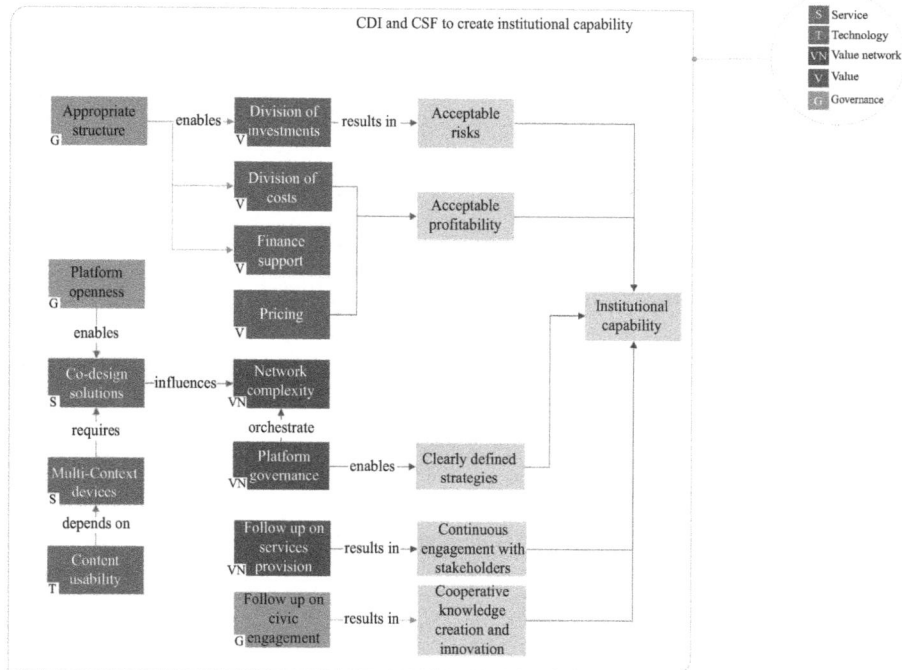

Figure 9.6   Creating institutional capability CDIs and CSFs

- Continuous Engagement with Stakeholders
- Cooperative Knowledge Creation and Innovation

9.4.5   Providing support to physical capability

The physical capability represents the city's endowment of hard infrastructure. The success and decline of cities depends upon the technological infrastructure adopted by cities' policy-makers [23]. This dependence is said to be directly proportional to the size of the city: the larger the city, the stronger the dependence. The provision of multi-context services (service domain) demands the co-design of solutions (service domain) that creates the first CSF **Development of Innovative Services**.

Interoperability (technology domain) which includes data, systems and services interoperability, alongside content reusability (technology domain), content usability (technology domain) and security mechanisms (technology domain), influences the QoS, that if acceptable and sufficient will result in the CSF **Increase Responsiveness of Systems** and **Sustain Knowledge Creation**. Such CSFs provides support to inter-connect the physical-digital realms of cities and end the *trial separation of bits and atoms* as suggested by Cash [144]. The latter is also supported by value elements such as trust and privacy (service domain) and accessibility (technology domain) which

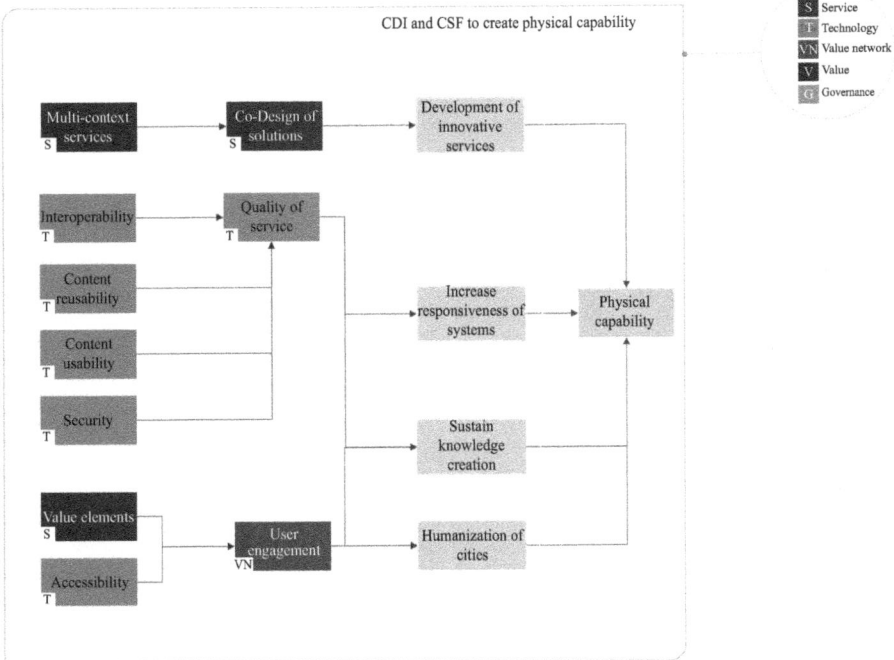

Figure 9.7   Creating physical capability CDIs and CSFs

will in turn enable users to engage more in public services. Research has demonstrated that in some cases the physical capability of cities can exacerbate inequality due to resource allocation, investment and maintenance being partial to perceived sources of *value creation* and *competitive articulation* in urban economies [12,43]. Such engagement and participation has the potential to address this issue by humanizing cities and create the fourth CSF **Humanization of Cities**. As a result, the public who was just an outsider before, now is part of the data processing system and might process, enrich, share, combine, input their own data (e.g. mobile phones, personal devices) and even create high-level streams of knowledge through crowd-sourcing and creation of applications.

The CSFs to provide support for creating physical capability are outlined below. High scores on these success factors will provide opportunities to improve the efficiency and the effectiveness of the urban space as well as the management and control of resources. A service that provides support for creating physical capability in the long run can be expected to result in a viable business model. Figure 9.7 illustrates the CSFs for creating physical capability.

- Development of Innovative Services
- Increase Responsiveness of Systems
- Sustain Knowledge Creation
- Humanization of Cities

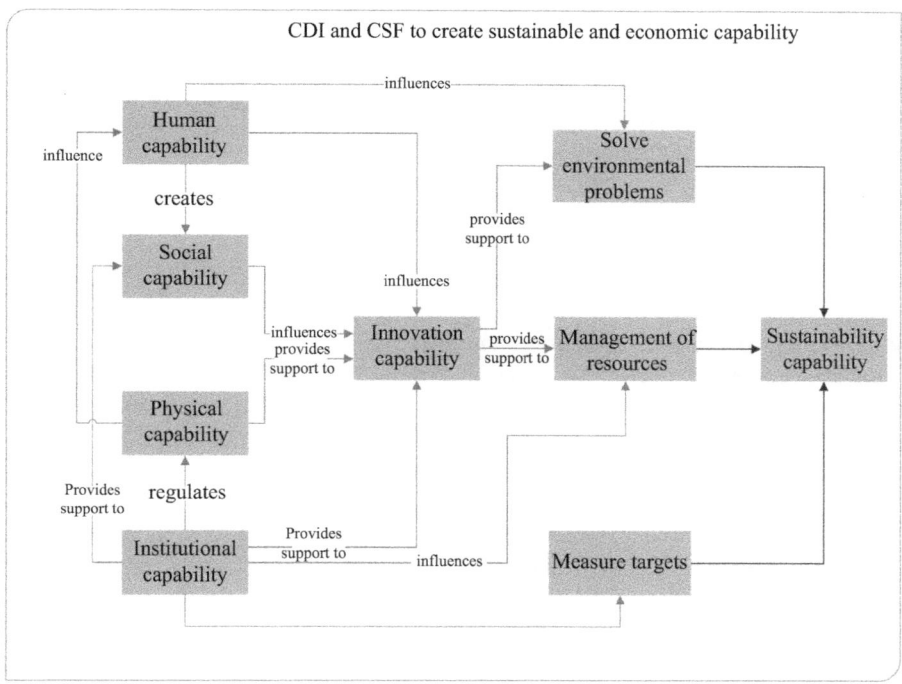

Figure 9.8   Creating sustainable and economic capability CDIs and CSFs

9.4.6   Providing support to sustainable and economic capability

The social and human capabilities, alongside the physical capability, have the potential to amplify local creativities, networking, experimentation and innovation. The city local system of innovation is enhanced by digital collaboration spaces, interactive tools and embedded systems. Consequently, city gains innovation capability, which is translated into increased competitiveness, a better environment, more jobs and wealth, workforce development and improvement in productivity [112,422].

As a main goal, cities are looking for intelligent solutions that will provide a high quality of life without excessive operating costs and enable them to evolve and appropriately react to environmental changes and recover from unpredictable situations while meeting ambitious targets for innovation and sustainability agendas. The need for a sustainable development is near the top of all cities' agendas [49–51]. For instance, Vodafone and Accenture's report "The Carbon Connections"[1] published in 2009 showed that the use of digital technologies, billions of mobile connections and M2M data transmission, could potentially save 2.4% of expected EU emissions

---

[1] https://www.vodafone.com/content/dam/vodafone-images/sustainability/downloads/carbon_connections.pdf

that is 113 million tonnes of CO_2, which is equivalent to saving 18% of UK emissions in 2008. This would save 43 billion in energy costs alone. As quality of life becomes an important source of competitive advantage, cities have to provide a clean, green and safe environment for their citizens. Planning, transport, finance and economic policies all need to reflect the environmental goals that a city sets for itself. On the other hand, citizens also need to be engaged in the development and implementation of environmental policies and be encouraged to take responsibility for the quality of the environment in which they live.

The data infrastructure of cities will trigger the sustainable development by releasing data that will enable cost-effective government transparency projects, improvement of services, creation of new and innovative services, stimulation of competitiveness, stimulation of knowledge developments including a more active participation and empowerment of citizens, and the re-use of information across multiple processes, systems, value chains and stakeholders. We argue that the CSF of Sustainable and Economic capabilities is **Sustainable Economic growth** of cities. And this CSF is enabled by the human, social, innovation, physical and institutional capabilities of cities. A data infrastructure that provides support for creating the capabilities of cities that will drive sustainable economic growth in the long run can be expected to result in a viable business model. Figure 9.8 illustrates the CSFs for Creating Sustainable and Economic Capability.

9.5   Summary

In the first section of this chapter, we discussed the relevance of service innovation in smart cities. We also discussed trends and drivers that stimulate the development of data infrastructure and defined our core concepts. Our focus is on the design of data infrastructures for smart cities and their underlying business models. At the moment, there is no common framework for data infrastructure design nor is there a common framework for designing a common reference architecture for data infrastructure and their business models.

In this chapter we have tried to develop the first framework part of the SMARTify approach. We have discussed the theoretical foundation for our business model approach, defining the business model concept as well as the core concepts and specifying the relevant concepts in the business models domains. We looked at the interdependencies within as well as between the domains and discussed the core concepts from a design perspective, as well as the CDIs. We also explained in detail the viability and feasibility of business models by looking into CSFs.

The CSFs represent how well the data infrastructure is providing support to the development of urban capabilities. We did not intend to rank the importance of urban capabilities but to treat them as equally important. Our CSFs were developed based on the constant need that decision makers have in understanding how their initiatives fit in the wider context of smart cities. We expect it will provide means for them to measure the impact of their initiative and how it has contributed to the realization of

Table 9.3 Mapping value CDIs with requirements

CDIs	Service	Technology	Value Network	Value	Governance
Clear Defined Target Group	Multi-context services				
Compelling Citizen's Driven Services	Value elements, reputation		Partner selection	Pricing	
Increasing Participation		Data/metadata quality, security, QoS, accessibility, usability		Follow up on civic engagement	
Acceptable QoS					
Increasing Sense of Community				Demonstration of impacts	Appropriate structure, Social Inclusion
Optimized Collective Intelligence					Appropriate structure, Social co-operation
Unobtrusive Participation in Urban Services	Value elements	Accessibility, crowd-sourcing			
Local Experimentation and Innovation				Finance support	Appropriate structure
Open Innovation	Co-design of solutions		Broaden partnership	Demonstration of impacts	Platform Openness
Innovation Transfer				Broaden partnership	Platform Openness
Continuous Innovation	Multi-context services	Interoperability, content usability, content reusability			Follow up on civic engagement, Follow up on Creativity

(Continues)

Table 9.3 (Continued)

CDIs	Service	Technology	Value Network	Value	Governance
People-driven Innovation	Multi-context services	Interoperability, content usability, content reusability	User engagement		
Acceptable Risks					Appropriate structure
Acceptable Profitability				Division of investments	
Clearly Defined Strategies	Co-design of solutions, Multi-context services	Content usability	Network complexity, platform governance	Division of costs, finance support, pricing	Platform Openness
Continuous Engagement with Stakeholders					Follow up on services provision
Cooperative Knowledge Creation and Innovation					Follow up on civic engagement
Development of Innovative Services	Multi-context services, co-design of solutions				
Increase Responsiveness of Systems		Interoperability, content usability, content reusability security, QoS			
Sustain Knowledge Creation		Interoperability, content usability, content reusability, security, QoS			
Humanize cities	Value elements	Accessibility	User engagement		

the smart cities vision. Ultimately, we wanted to understand what the effects are of specific choices and what really contributes to the viability and feasibility of a data infrastructure and its business models. A negative assessment of certain CSFs implies that there will be bottlenecks in the business model's viability and that CDIs related to such CSFs should be redesigned. Table 9.3 provides an overview of the CSFs and CDIs that should be re-examined when a CSF is evaluated negatively.

Chapter 10

Applied data infrastructures assessment

The assessment of the business models viability is performed by systematically clustering the critical design issues (CDIs) and using them to assess and balance the requirement specifications elicited in the business models analysis. In the case any requirement specification negatively impacts a critical design issue, a requirements trade-off analysis must be carried out. For instance, consider the simplistic example in which a data infrastructure must support the federation of data from external data sources and at the same time satisfy pre-defined CDIs, such as Target Users (service), User engagement (service), Interoperability (technology), and Broaden Partnership (organization). On the one hand, increased content targets engage users with the data platform as well as increase the partnership with external data providers (contributing partners). On the other hand, federating data from other datasets significantly compromises data interoperability and requires the implementation of several mechanisms to mitigate semantic mismatch. Based on such arguments, this requirement should not be satisfied at the moment and revisited at a later stage when circumstances change.

In the next sections, we provide examples of how to trade off requirements against the CDIs, and how to validate business models against the pre-defined critical success factors (CSFs).

10.1 Open data infrastructure case study

10.1.1 Evaluation with CDIs

Tables 10.1–10.4 present the mapping of requirements with the CDIs of the service, technology, value network, value and governance design for the Open Data Infrastructure case study.

In the value network domain, providing mechanisms for data providers to manage their resources and enable users to give feedback on them contributes to the Broaden Partnership CDI. Data regulators may reduce the Network Complexity by setting up regulations, policies and standards to be put in place.

10.1.2 Evaluation with CSFs

A negative assessment of certain CSFs implies that there will be bottlenecks in the business model's viability, and that CDIs related to such CSFs should be redesigned.

Table 10.1 Mapping services CDIs with service design requirements

Requirements	Targeting	Creating value elements	Multi-context services	Co-design of solutions	Reputation	User engagement
FREQ.1	+			+		+
FREQ.2	+	+			+	
FREQ.3	+				+	+
FREQ.4	+	+			+	
FREQ.5	+			+	+	+
FREQ.6	+			+	+	+
FREQ.7	+	+			+	+
FREQ.8		+			+	+
FREQ.9	+		+		+	+
FREQ.10	+	+			+	+
FREQ.11	+				+	
FREQ.12		+			+	
FREQ.13		+	+		+	
FREQ.14	+				+	+
FREQ.15	+	+			+	
NFREQ.1		+			+	+
NFREQ.2		+			+	+
NFREQ.3	+	+			+	+
NFREQ.4	+	+			+	+
NFREQ.5	+	+			+	+
NFREQ.6	+	+			+	+
NFREQ.7	+	+			+	+
NFREQ.8		+			+	+

Table 10.5 provides an overview of the assessment of the CSFs in the Smart City Platform case study. When evaluated negatively, the CDIs that influence the CSF should be re-evaluated.

The CSF **compelling citizen's driven services** and **innovation transfer** were evaluated negatively. This is due to the fact that the platform will only provide data and no data manipulation services. This functionality could become a competitive advantage over other available platforms, would facilitate innovation transfer and put the Smart City at the forefront of smart cities data infrastructure initiatives. Besides, it should create greater opportunities for citizens and developers to create new products and services which will enable economic development in the Smart City. From discussions with decision makers, those are not important CDIs that will hinder the development and immediate success of the Smart City Platform. We recommend that

Table 10.2 Mapping technology CDIs with technology design requirements

Requirements	Data/metadata quality	Security	Quality of service	Usability	Content usability	Content reusability	Content interoperability	Content accessibility
FREQ.20	+	+						
FREQ.17	+	+						
FREQ.18					+	+	+	
FREQ.19		+						
FREQ.20		+						
FREQ.21		+						
FREQ.22		+						
FREQ.23	+				+	+		
FREQ.24				+	+	+		
FREQ.25		+						+
FREQ.26		+						
FREQ.27		+		+				
FREQ.28	+	+						
FREQ.29	+							
FREQ.30	+	+						
FREQ.31			+					
FREQ.32			+					
FREQ.33	+	+			+			
FREQ.34			+					
FREQ.35		+	+					
FREQ.36				+	+	+	+	+
FREQ.37								+
FREQ.38		+	+					
FREQ.39				+				+
NFREQ.9		+	+					

at this stage of the data infrastructure development, the Smart City launches a v1.0 platform without the capabilities of data services and commercial data, however, consider the extension of the platform to a v2.0 which would accommodate such advanced requirements.

10.1.3 Robustness check

The SMARTify approach delivered a complete business model which highlighted several business model *blind spots* that were not identified in the original documentation of the Smart City Platform. For instance, our approach provided a comprehensive list of requirements, complete identification of stakeholders and their expectations and

Table 10.3   Mapping technology CDIs with technology design requirements

Requirements	Data/metadata quality	Security	Quality of service	Usability	Content usability	Content reusability	Content interoperability	Content accessibility
NFREQ.10			+					
NFREQ.11			+					
NFREQ.12				+	+	+		+
NFREQ.13		+	+					
NFREQ.14		+	+					
NFREQ.15			+				+	
NFREQ.16		+	+					
NFREQ.17		+	+					

Table 10.4   Mapping value network, value and governance CDIs with value network, value and governance design requirements

Requirements	Broaden Partnership	Network Complexity	Partner Selection	Demonstration of Impacts	Appropriate Structure	Platform Openness	Follow up on civic engagement	Follow up on services provision
FREQ.40	+							
FREQ.41	+							
FREQ.42	+							
FREQ.43	+							
FREQ.44	+		+					
FREQ.45				+				
FREQ.46				+				
FREQ.47								+
FREQ.48							+	
FREQ.49					+			
FREQ.50					+	+		

Table 10.5 Open Data Infrastructure CSFs assessment

Capability	CFS	Status	Assessment
Human	**Clearly Defined Target Group**	Positive	The service focuses on providing high-quality open data to the stakeholders of smart cities.
	Compelling citizen's driven services	Partially	Through the provision of trustable, ease of use and high-quality open data the platform is able to provide some services that will satisfy the demand.
	Increasing Participation	Positive	By providing easy to use and accessible interfaces, and data consumption channels the platform will increase the engagement of users.
	Acceptable Quality of Service	Positive	Scalability of the platform can be achieved by relying on cloud computing infrastructures.
Social	Increasing Sense of Community	Positive	The effective use and re-use of open data provided by the platform can give rise to user-oriented applications which strengthen community development potential.
	Optimized Collective Knowledge	Positive	The platform provides data and services that support people-driven innovation by introducing principles of standardization, openness and removal of barriers to data access.
	Unobtrusive participation in urban services	Positive	The increased accessibility (ease of use and standard interfaces) alongside value elements trust, privacy preserving) and feedback collection makes it easier to obtain user-generated content.
Innovation	Local Experimentation and Innovation	Negative	There is a limited amount of services provided by the platform (data query and consumption) that does not support the manipulation of data that can be interoperable and used in research, smart spaces and emerging technologies to amplify local creativity, experimentation and innovation.
	Open Innovation	Partially	The platform does not provide means to external people to collaborate in the provision of services and data, but just alliances to define licensing and standardization schemes.
	Innovation Transfer	Negative	There is no provision of personalized tools nor openness strategies that supports partnerships in data and services provision which would facilitate innovation transfer in the urban environment.
	Continuous enables Innovation	Positive	The mechanisms for platform owners to follow up on civic engagement and on creativity data infrastructure providers to understand which kinds of innovations are being created and what kind of data and services is still needed.

(Continues)

Table 10.5 (Continued)

Capability	CFS	Status	Assessment
	People-driven innovation	Positive	Multi-context services give users access to data and services that facilitates the emergence of smart clusters and innovation ecosystems of entrepreneurs and people.
Institutional	Acceptable Risks	Positive	Platform owners will have access to mechanisms that enables them to define terms and conditions for services and data provision.
	Acceptable profitability	N/A	N/A
Institutional	Local Experimentation and Innovation	Negative	There is a limited amount of services provided by the platform (data query and consumption) that does not support the manipulation of data that can be interoperable and used in research, smart spaces and emerging technologies to amplify local creativity, experimentation and innovation.
	Clear Defined Strategies	Positive	Although it is not a functionality of the platform, in the business models it is explicitly required that stakeholders should be engaged decision making (e.g. standards, licensing, profitability, privacy agreements).
	Continuous engagement with stakeholders	Positive	Platform owners will have mechanisms to collect information on services and data provision, as well as feedback from user, which enables them to track how data is being used and their quality level.
Physical	Development of Innovative Services	Positive	The provision of multi-context open data is enabled by the co-design of innovative solutions.
	Increase responsiveness of systems	Positive	The release of open data has the potential to improve the service level of urban systems.
	Sustain Knowledge Creation	Positive	Interoperable data, alongside Content Reusability and usability and the security mechanisms of the platform, results in acceptable services that will be used for knowledge creation.
	Humanize Cities	Positive	The platform Value Elements, such as trust and privacy, and accessibility and usability facilitate the engagement and participation of humans in the data processing and knowledge creation of cities.

resources. Furthermore, the CDIs helped to validate the requirements needed to arrive at sensible business model design choices and provide support to the development of urban capabilities. Our approach has delivered a reference architecture that can be used to make an informed decision when choosing among the vast number of open data platforms provided in the market. The SMARTify approach helped preparing and conducting brainstorming and interview sessions, and gathering requirements for the data infrastructure.

The service domain opened up the discussion regarding limiting the platform to open data provision at its launch and then at a later stage to expand the capabilities of the platform by incorporating data commercialization tools and advanced data processing services. It also became clear that there is a need for a more concrete and elaborated description of the service, and that this requires more interaction with and knowledge about the target markets. The organization domain draws attention to the role of data regulators and the need for a new or existing actor to take up this role. By introducing data regulators in the value network, the need for data publishers to define standardization options was removed. In addition, the organization domain makes it clear that the initiative requires a complex organizational network that depends on the cooperation of multiple stakeholders in the decision-making process.

Requirements concerning the data architectural features of the data were not identified using the current approach. The comprehensive technology domain allowed practitioners to gather critical requirements that will enable the platform to serve the growing demand for high-quality open data. The feedback from practitioners was very positive and they confirmed the approach address relevant issues in a systematic way, is complete and comprehensible to be used in the industry environment. They were capable of identifying the technical and non-technical components of the data infrastructure in an integrated manner and also understand how this strategy fits in the context of smart cities and the government agenda. However, the practitioners pointed out that the business models do not incorporate methods to facilitate the financial aspects of the data infrastructure, the management of investments and estimation of costs. We understand the importance of this shortcoming and we provide a discussion on the available tools that can support the modelling of financial aspects in data infrastructure design.

10.2   London Data Infrastructure case study

10.2.1   Evaluation with CDIs

Table 10.6 presents the mapping of requirements with the CDIs of the service domain.

The data infrastructure owner should make trade-offs in order to accommodate the users' requirements as well as align them to the data infrastructure governance and technical architecture. Table 10.7 presents the mapping of technology requirements with Technology CDIs. In the technology domain, there is a clear trade-off in between providing linked data (FREQ.43) and ensuring the CDI Quality of service. This is due to the fact that enriched data tends to be very large and wordy which may hurt

Table 10.6   Mapping services CDIs with service design requirements

REQUIREMENTS	Targeting	Creating Value Elements	Multi-context services	Co-Design of Solutions	Reputation	User engagement
FREQ.1	+		+	+		+
FREQ.2	+	+				
FREQ.3	+	+			+	
FREQ.4	+	+				
FREQ.5	+		+			
FREQ.6	+		+			
FREQ.7	+			+		
FREQ.8	+	+	+			
FREQ.9	+		+			
FREQ.10	+		+		+	
FREQ.11		+			+	
FREQ.12	+		+			+
FREQ.13	+		+			+
FREQ.14	+		+			
FREQ.15			+			
FREQ.16			+			
FREQ.17	+		+			
FREQ.18	+	+			+	
FREQ.19	+	+			+	
FREQ.20			+		+	
FREQ.21	+				+	+
FREQ.22			+			
FREQ.23			+		+	
FREQ.24			+			
FREQ.25			+			
FREQ.26	+	+				+
FREQ.27	+					+
FREQ.28	+					+
FREQ.29	+				+	+
FREQ.30	+	+				+
FREQ.31	+		+			+
FREQ.32			+			
FREQ.33	+	+	+		+	+
FREQ.34	+			+	+	
FREQ.35				+	+	
FREQ.36	+				+	
FREQ.37					+	
NFREQ.1						
NFREQ.2						+

(Continues)

Table 10.6   (Continued)

REQUIREMENTS	Targeting	Creating Value Elements	Multi-context services	Co-Design of Solutions	Reputation	User engagement
NFREQ.3						
NFREQ.4						
NFREQ.5		+				+
NFREQ.6		+				++
NFREQ.7						++
NFREQ.8					+	
NFREQ.9					+	+

Table 10.7   *Mapping technology CDIs with technology design requirements*

REQUIREMENTS	Data/metadata quality	Security	Quality of service	Usability	Content usability	Content reusability	Content interoperability	Content accessibility
FREQ.38		+						+-
FREQ.39	+		+-		+	+	+	
FREQ.40		+						
FREQ.41	+	+			+	+		
FREQ.42		+						
FREQ.43	+	+	+					
FREQ.44	+				+	+	+	
FREQ.45	+				+	+	+	
FREQ.46		+						
FREQ.47	+			+	+	+		
FREQ.48				+	+	+	+	+
FREQ.49					+	+	+	+
FREQ.50					+	+		+
FREQ.51	+				+	+	+	
FREQ.52		+						

(Continues)

Table 10.7 *(Continued)*

REQUIREMENTS	Data/metadata quality	Security	Quality of service	Usability	Content usability	Content reusability	Content interoperability	Content accessibility
FREQ.53	+	+		+			+	
FREQ.54		+						
FREQ.55		+						
FREQ.56		+						
FREQ.57		+		+				+
FREQ.58		+						
FREQ.59		+						
FREQ.60	+	+						
FREQ.61	+	+						
FREQ.62	+	+						
FREQ.63	+	+					+	
FREQ.64	+	+	+				+	
FREQ.65	+	+	+				+	
FREQ.66	+	+					+	
FREQ.67		+	+					
FREQ.68	+	+					+	
FREQ.69	+	+						
FREQ.70	+	+						+
FREQ.71		+	+					
FREQ.72	+	+						
FREQ.73		+	+					
FREQ.74		+	+					
FREQ.75	+	+	+					
FREQ.76		+	+					
FREQ.77				+				+
FREQ.78	+	+	+					
FREQ.79				+	+	+		
FREQ.80	+	+					+	
NFREQ.10			+					
NFREQ.11			+		+	+		+
NFREQ.12			+		+	+		+
NFREQ.13		+	+					
NFREQ.14		+	+	+				
NFREQ.15		+	+					
NFREQ.16		+	+				+	+
NFREQ.17		+	+					+
NFREQ.18		+	+					
NFREQ.19		+	+					
NFREQ.20		+	+					

the performance of the data infrastructure. It should be taken into consideration when choosing the backbone infrastructure of the data infrastructure. Providing mechanisms to verify data quality and completeness (FREQ.55) also puts a great demand for performance and bandwidth and should be also taken into consideration.

Table 10.8 presents the mapping of value network, value and governance CDIs with value network, value and governance design requirements. In the organization domain, there is a clear trade-off in between enable data facilitators to extend the services of the data infrastructure (FREQ.81) and managing the CDIs Network Complexity. It would raise organizational issues, with a larger number of organizations becoming involved in the decision-making process. There is a clear trade-off between choosing a limited target group leading to reduced benefits from the service for end-users on the one hand and choosing a broader target group leading to higher technological and organizational complexity. NFREQ.21 (Extensibility & Maintainability) also affects the Network Complexity. Appropriate structures in the governance domain may be able to handle such issues.

In the value domain, there is a clear trade-off in between enable application of taxes on the commercialization of services and data (FREQ.92) and pricing and division of costs and revenue. Data and service providers should agree on an acceptable pricing scheme so that when tax is applied to the service/purchase the final price is still appealing to users. Data infrastructure and data providers and complementors should make an agreement on acceptable pricing so that there is no abuse that would affect the engagement of users with the data infrastructure.

After the requirements mapping (Quick Scan) has been carried out, there are five domain-related designs that together make up the initial business model, and they are a description of the service and the intended value proposition, a value network, a technical architecture, value and governance model. It is likely that the requirements in the five domains affect each other. For instance, can the technical architecture deliver the value proposition? Is the network of stakeholders able to deliver services and their value proposition? Is the platform governance orchestrating the value network? We have mapped all the requirements gathered to the CDIs of the five domains to check for conflicting requirements and obstacles. To increase the balance between the domains, we have made few trade-offs to guarantee the business model outline is aligned with the value proposition. In this case study, we found that some requirements contribute negatively with the CDIs of the business models. For instance, one of the conflicting requirements is sharing user's personal information with the providers of data and services (FREQ and the User Engagement CDI. A trade-off is to only share the anonymized general user's data with the services providers of the platform. Tables 10.9–10.12 present some of the conflicting requirements and CDIs found in this study, and the trade-off and criteria adopted to solve each conflict.

10.2.2 Evaluation with CSFs

A negative assessment of certain CSFs implies that there will be bottlenecks in the business model's viability, and that CDIs related to such CSFs should be redesigned. Table 10.13 provides an overview of the assessment of the CSFs in the London Data

Table 10.8   Mapping value network, value and governance CDIs with value network, value and governance design requirements

REQUIREMENTS	Broaden Partnership	Network Complexity	Partner Selection	Demonstration of Impacts	Pricing	Division of Investments and Risks	Division of Costs and Revenue	Appropriate Structure	Platform Openness	Follow up on civic engagement	Follow up on services provision
FREQ.81	+	+-									
FREQ.82	+		+								
FREQ.83	+										
FREQ.84	+										
FREQ.85	+										
FREQ.86	+										
FREQ.87	+		+								
FREQ.88	+			+							
FREQ.89					+		+				
FREQ.90					+		+				
FREQ.91					+		+				
FREQ.92					+-	+	+-				
FREQ.93				+		+	+				
FREQ.94				+		+					
FREQ.95					+	+	+				
FREQ.96				+			+				
FREQ.97				+			+				
FREQ.98											+
FREQ.99										+	
FREQ.100								+			
FREQ.101								+			
FREQ.102								+	+		
FREQ.103								+	+		
FREQ.104								+	+		
NFREQ.21	+	+-									
NFREQ.22	+										
NFREQ.23	+										
NFREQ.24				+							
NFREQ.25							+				

Table 10.9 Requirements trade-off: interoperability

Design Issue	Interoperability
Domain	*Technology*
Balancing requirements	FREQ.49
Solution	The platform should encourage the publication of a great variety and quantity of data as it can be used to create informed decisions and innovative services in the city. Federating data from other datasets may compromise the interoperability of data, hence data federation should not be allowed at this stage.
Criteria	A great range of interoperable data is desirable but federating data may compromise the interoperability of the data and should be avoided.

Table 10.10 Requirements trade-off: quality of service

Design Issue	Quality of service
Domain	*Technology*
Balancing requirements	FREQ.39, FREQ.44, FREQ.45
Solution	Graph data annotation is necessary to ensure the security and trustworthiness of the data as well as guarantee data ownership and usage rights. Data manipulation can facilitate the usability and reusability of data by users and should be provided.
Criteria	Include scalability as a CDI in the Technology domain to ensure the platform can support linked data modelling and accommodate data manipulation services.

Table 10.11 Requirements trade-off: network complexity

Design Issue	Network complexity
Domain	*Value Network*
Balancing requirements	FREQ.81, FREQ.82
Solution	These requirements increase the complexity of the network by requiring complementors to join the network of actors and extend the services of the platform. However, the value perceived by the provision of extended services is higher than the complexity of the network which can be addressed by adopting appropriate strategies.
Criteria	The provision of extended services should be encouraged.

Table 10.12   Requirements trade-off: pricing

Design Issue	Pricing
Domain	*Value*
Balancing requirements	FREQ.92
Solution	Although taxes may increase the cost of data, appropriate structure requirements ensures there is an acceptable profitability.
Criteria	The platform should comply with the any laws regarding financial transactions.

Store case. When evaluated negatively, the CDIs that influence the CSF should be re-evaluated in Step 3.

10.2.2.1   Robustness check

We believe that this case study proved to be a valid evaluation effort. The guidelines and techniques present in SMARTify offered appropriate support to conduct the evaluation and design of business models for data infrastructures. In this chapter, we presented the use of the SMARTify model and method to develop the London digital infrastructure business model starting from an existing infrastructure providing limited open data.

The SMARTify approach helped preparing and conducting interviews, and gathering requirements for the data infrastructure. The presentation of the results in a feedback session to practitioners gave the impression that the approach addresses relevant issues in a systematic way. We experienced that going through each domain separately and merging them at a later stage provided more insights and topics for discussion. The development of the business model resulted in a change from a focus on the technology toward an overall picture that includes the service, organization and value and governance domains. It resulted in a rethinking of services, technology and governance choices that are in place at the existing data infrastructure of London (London Datastore). The service domain opened up the discussion regarding a use of additional services provided by complementors and proprietary data to serve other target groups. It also became clear that there is a need for a more concrete and elaborated description of the service and technology, and that this requires more interaction with and knowledge about technologies and services available. It required sacrificing some performance for future flexibility (providing linked data to increase content usability and reusability).

The organization domain draws attention to the new role of services complementors in providing services that would be difficult to develop in-house. It emphasizes the need of developing data infrastructures in which its governance strategy supports multiple actors. It makes clear that the initiative requires a complex organizational network that depends on the cooperation of multiple stakeholders with different interests and capabilities. While in general there is agreement that data and services could be commercialized, the value domain showed that the revenues and costs are not yet

Table 10.13 London data infrastructure CSFs assessment

Capability	CFS	Status	Assessment
Human	Clearly Defined Target Group	Positive	The service focuses on providing high-quality open and commercial data and services to the stakeholders of smart cities.
	Compelling citizen's driven services	Positive	Through the provision of trustable, interoperable data and engaging services, the data infrastructure is able to provide services that will satisfy the demand.
	Increasing Participation	Positive	The use of open, standard and accessible interfaces and data consumption channels in the data infrastructure will increase the engagement of users.
	Acceptable Quality of Service	Negative	Scalability issues pose a threat to the validity of the business models. Scalability should be regarded as an important CDI.
Social	Clearly Defined Target Group	Positive	The service focuses on providing high-quality open and commercial data and services to the stakeholders of smart cities.
	Increasing Sense of Community	Positive	The re-use of cross domain integrated city data provided by the data infrastructure can significantly increase the development of user-oriented applications and Smart City services which strengthen community development potential.
	Optimized Collective Knowledge	Positive	The data infrastructure provides data and services that supports people-driven innovation by introducing principles of standardization, openness and removal of barriers to data access.
	Unobtrusive participation in urban services	Positive	The increased accessibility (ease of use and standard interfaces) alongside value elements (e.g. trust, privacy preserving) and feedback collection makes it easier to obtain user-generated content and increased engagement.
Innovation	Local Experimentation and Innovation	Positive	The services and data provided by the data infrastructures supports the creation of innovative services and research, smart spaces and emerging technologies which amplifies local creativeness, experimentation and innovation.
	Open Innovation	Positive	The platform provides means to external people to collaborate in the provision of services and data, and alliances to define licensing, policies, financial models and standardization schemes.
	Innovation Transfer	Positive	Interactive and personalized tools and the openness of the platform in supporting partnerships in data and services facilitate innovation transfer in the urban environment.

(Continues)

Table 10.13 (Continued)

Capability	CFS	Status	Assessment
	Continuous Innovation	Positive	The mechanisms for platform owners to follow up on civic engagement and on creativity enable data infrastructure providers to understand its role in the creation of innovative services and the market demand for city data.
	People-driven innovation	Positive	Multi-context services give users access to data and services that facilitates the emergence of smart clusters and innovation ecosystems of entrepreneurs and businesses.
Institutional	Acceptable Risks	Positive	Platform owners will have access to mechanisms that enables them to define terms and conditions for services and data provision, including division of investments and costs, revenues, and financial models.
	Acceptable profitability	Positive	Platform owners will have access to mechanisms that enables them to regulate the profitability and taxes of financial transactions in the data infrastructure.
Institutional	Clear Defined Strategies	Positive	Although it is not a functionality of the data infrastructure, the business models explicitly require that stakeholders should be engaged in the design of the data infrastructure and decision making (e.g. standards, licensing, profitability, privacy agreements).
	Continuous engagement with stakeholders	Positive	Platform owners will have mechanisms to collect information on services and data provision, as well as feedback from user, which enables them to track how services are used and their quality level.
Physical	Development of Innovative Services	Positive	The provision of multi-context data and services supports the development of innovative solutions.
	Increase responsiveness of systems	Negative	Scalability issues pose a threat to the validity of the business models. Scalability should be regarded as a CDI.
	Sustain Knowledge Creation	Positive	Interoperable data, alongside content reusability and usability and the security mechanisms of the data infrastructure, results in acceptable services that will be used for knowledge creation and innovation.
	Humanize Cities	Positive	The data infrastructure Value Elements, such as trust and privacy, and accessibility and usability facilitate the engagement and participation of humans in the data processing and knowledge creation of cities.

very clear. Moreover, the distribution of the revenues and costs among the various actors remains an important issue that needs to be resolved. The reference architecture serves as a holistic overview of the platform to be developed or outsourced from open data platform vendors. Our approach resulted in rethinking the technology choices that were made with regard to the existing data infrastructure. The existing infrastructure is a high-end solution that is not suitable to provide multi-context data and services, and as a result, a good portion of the target market is left out. The cost of deploying such data infrastructure must be considered in more detail with all the members of the value network.

10.3 Complementary tools and techniques

10.3.1 Volumetric analysis of city data

The following sections give examples of volumetric data estimation of city data. These datasets were used to identify the architecture features and flows of city data. We describe the following transport datasets: Smart Parking, Parking Meters, Bike Hiring Scheme, Tube line feeds, Traffic Disruption feeds, transport smart cards journey data (e.g. Oyster (UK), Pasmo (Japan)) and Taxi GPS Traces. Figure 10.1 illustrates these data sets and their respective flows in the urban environment.

10.3.1.1 Transport data

SMART PARKING: 80,000 units of smart parking sensors for a region with the size of Greater London, all with sensing rate of 1–4 times/hour. The sensing rate of 4 times/hour would occur during peak time which lasts around 7 hours (9:00-16:00), and the sensing rate of 1 time/hour would occur during non-peak time, which represent 17 hours in total. We assume that each sensing would generate 102 bytes of data. The data assumptions are shown in Table 10.14.

BICYCLE HIRING: 8,000 'Boris Bikes' and 570 docking stations in the Barclays Cycle Hire scheme. These bikes have been used for around 18,500 journeys per day and have made over 10 million journeys. We assume that each hire transaction generates 50 bytes of data. The data assumptions are shown in Table 10.15.

UNDERGROUND STATUS: TfL provides live feeds containing an XML representation of all London tube line status. The data present in the live feeds is shown in Table 10.16.

TRAFFIC DISRUPTION: TfL provides live feeds containing an XML representation of all traffic disruptions within Greater London. The data present in the live feeds is shown in Table 10.17.

SMART CARDS: Every week, the TfL processes 57 million journeys data (including London Underground, National Rail, DLR, Overground, Buses, etc.). Around 67% of these journeys are made using oyster cards, representing 2,669,338 daily journeys which generate data on the TfL system. We assume each Oyster Card transaction feeds the system with basic information of the user's journey as shown in Table 10.18.

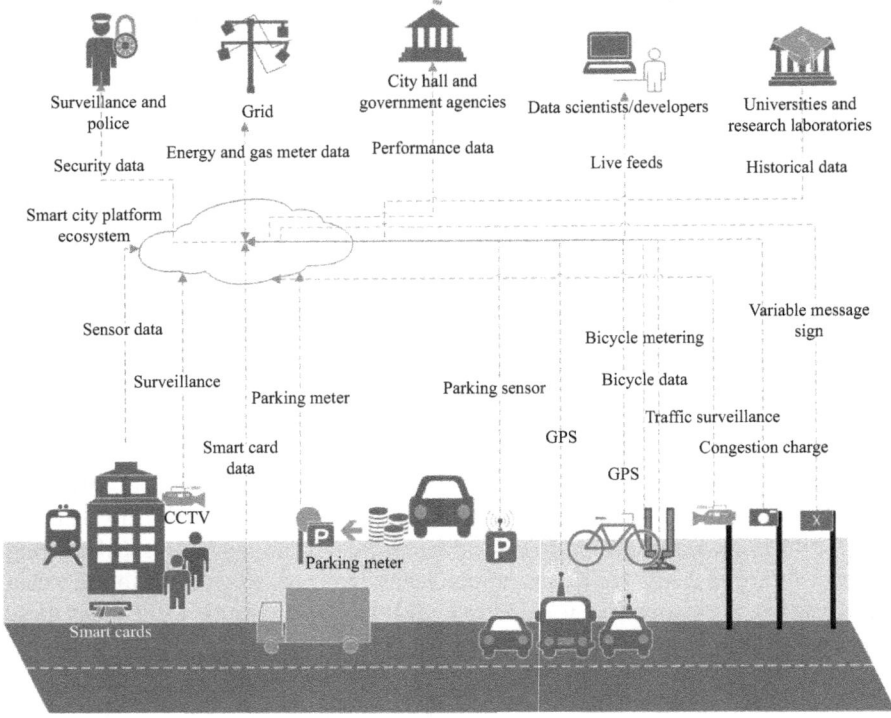

Figure 10.1   Transport datasets

Table 10.14   Smart parking data

Field	Instance	Description	Bytes
Date	20120910	Sensing date	4
Time	113254	Sensing time	3
ID	1231X	Sensor unique ID	31
Status	F	Availability (T – Vacant/F – Busy)	2
Longitude	114.874565	Degree	31
Latitude	22.784498	Degree	31
TOTAL			102

TAXI GPS DATA: There are nearly 23,000 licensed taxis in London and 13,587 in New York City. We assume that everyday each taxi generates 1 GPS packet of 83 bytes every 10 s, which corresponds to approximately 40 Mb/h. This data can be collected for better taxi booking and planning services. The data assumptions are shown in Table 10.19.

Table 10.15   Bicycle hiring data

Field	Instance	Description	Bytes
Date	20120910	Hiring date	4
Time	113254	Hiring time	3
Bike_ID	24	Bicycle unique ID	4
Authcode	FFSDF454F6SD	Payment Authorization Code	31
Dock_St_ID	45	Docking station and unique ID	4
Client ID	845138	Client unique ID	4
TOTAL			50

Table 10.16   Underground status update

Field	Instance	Description	Bytes
Date*	20120910	Status date	4
Time*	113254	Status time	3
Line Status ID	9	Line unique ID	4
Status Details	Due to an obstruction on the track in the Westminster area	A description of the status of the line if the status is not normal otherwise this will be blank	183
Line ID	11	A code representing the line	4
Name	Metropolitan	The line name	31
Status ID	DS	A numeric code representing the status of the line	4
CssClass	Disrupted service	A text code representing the general status of the line	31
Description	Part suspended	A description of the status of the line	31
Is Active	True	A Boolean indicating if the status shown is active	4
Status Type	ID1	A code representing the status type the service is checking	4
Description	Tube status	A description of the status type the service is checking	4
TOTAL			307

10.3.1.2   Smart metering and sensing data

In this section, we describe the following smart metering and sensing datasets: water metering, water sensing, energy metering and smart grid, gas metering, smart appliances, smart buildings and smart street lights. Figure 10.2 illustrates these data sets on the urban environment.

Table 10.17 Traffic disruption data

Field	Instance	Description	Bytes
Disruption ID	1449	Unique disruption identifier	4
Status	Active	The status of the disruption	4
Severity	Severe	The impact of the disruption on traffic	4
Level of interest	High	Level of potential impact on traffic operations of the disruption	4
Category	Accident	This describes the nature of disruption	4
Start time	2013-02-05T16:33:00Z	The date and time which the disruption started	8
Location	Blackfriars Road (Southwark)	Main road name/number (borough) of the disruption	31
Corridor	Farringdon Cross Route	Road corridors affected by the disruption	31
Comments	Northbound direction	Full text of comments describing the disruption	183
Current update	Lane one (of three) is currently restricted	Text of the most recent update from the LSTCC on the state of the disruption	31
Remark time	2013-05-02T15:44:39Z	The time when the last current update was recorded	8
LastModTime	2013-05-02T15:44:39Z	Time when the last change was made to the database entry for the disruption	8
Coordinates EN	531,650.528, 18,0246.667	Easting and northing grid references	31
Coordinates LL	−0.104486, 51.505755	Longitude and latitude coordinates	31
Name	Blackfriars Road	Street name	31
Closure	Open	Type of road closure	3
Directions	North Bound	Set of directions on the road	31
Toid	4000000030239261	A 16-digit unique integer identifying a road link	31
Line coordinates EN	531,650.528, 18,0246.667	Co-ordinates of a line – easting and northing grid references	31
Line coordinates LL	−0.104486, 51.505755	Co-ordinates of a line – longitude and latitude coordinates	31
TOTAL			307

Table 10.18   Smart cards

Field	Instance	Description	Bytes
downo	3	A number between 1 and 7, 1 being Sunday, 2 being Monday etc.	1
daytype	Tue Sun–Sat	4	
SubSystem	LUL The mode(s) of the journey. LUL – London Underground, NR – National Rail, LTB – London Buses, DLR- Docklands Light Railway, LRC – London Overground, TRAM – Croydon Tram	4	
StartStn	Unstarted	Station the journey started at	31
EndStation	Kings Cross M	Station the journey ended at	31
EntTime	0	Entry time of the journey in minutes after midnight	4
EntTimeHHMM	00:00	Entry time in HH:MM text format	3
ExTime	633	Exit time of the journey in minutes after midnight	4
EXTimeHHMM	10:33	Exit time in HH:MM text format	3
ZVPPT	Z0104	Zones of Oyster Season ticket, if used	6
JNYTYP	TKT	Product types involved in the journey. PPY – Pure PAYG, TKT – Pure Oyster Season, MIXED – Combined PAYG and Oyster Season	4
DailyCapping	N	It shows as Y when PAYG journey was capped	1
Ffare	0	Full PAYG Fare before any discounts	1
Dfare	0	PAYG Fare after usage based discounts	1
RouteID	XX	The Route Number of the Bus, if a Bus has been boarded	4
FinalProduct	LUL	Travelcard 7 Day Combined Product Description used for journey	31
TOTAL			133

ENERGY AND GAS SMART METERING: Meters will be read every 15 min (96 times/day), and every energy meter generates 360 bytes/reading. In this analysis it was assumed that all London households, businesses and public buildings will have a smart meter by the end of the decade. For business and public building it was assumed there will be at least one energy and gas smart meters within its premises. The data assumptions are shown in Table 10.20.

Table 10.19 Taxi GPS data

Field	Instance	Description	Bytes
Date	20120910	Packet date	4
Time	113254	Packet time	3
Taxi_ID	24	Taxi unique ID	4
Company	XX	Company code	4
Speed	45	Km/h	4
Status	A Status (A-With client/B- Free)	2	
Longitude	114.874565	Degree	31
Latitude	22.784498	Degree	31
TOTAL			83

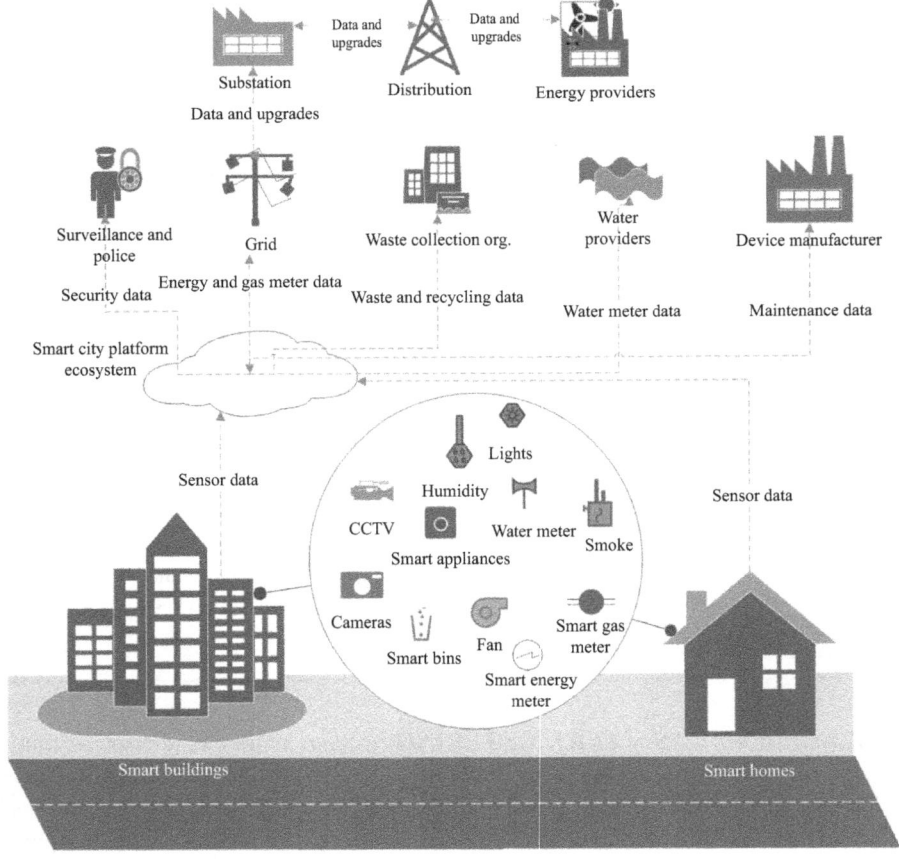

Figure 10.2 Smart metering and sensing data

Table 10.20   Energy and gas smart metering

Field	Instance	Description	Bytes
Date	20120910	Measurement date	4
Time	113254	Measurement time	3
Readings	–	Readings on the consumption, voltage, etc.	349
Meter_ID	24	meter unique ID	4
TOTAL			360

Table 10.21   Smart buildings

Field	Instance	Description	Bytes
Date	20120910	Measurement date	4
Time	113254	Measurement time	3
Readings	–	Readings on the temperature/ humidity/lights condition, etc.	39
Sensor_ID	12	Sensor unique ID	4
TOTAL			50

SMART BUILDINGS: All businesses and public buildings will smartly monitor air, temperature and light conditions. In this case, these conditions will be monitored every 5 min and generate 50 bytes/read. In this estimation, we assumed that a typical smart building have 250 sensors within its premises [425]. The data assumptions are shown in Table 10.21.

HOME AREA NETWORK (HAN): All Greater London's households have seven smart appliances (e.g. washing machine, smart dish washer, and smart refrigerator), each generating 400 bytes/day. The data assumptions are shown in Table 10.22.

SMART WATER METERING: All London households, businesses and public buildings will have a smart meter by the end of the decade. For business and public building, we assume that there will be at least one water smart meters within its premises. Water meters are read every 15 min (96 times/day), generating 260 bytes of data/read. The data assumptions are shown in Table 10.23.

WATER SENSING: We have assumed that all London households, businesses and public buildings will have a smart meter by the end of the decade. For business and public building, we assume that there will be at least one water smart meter within its premises. Water meters are read every 15 min (96 times/day), generating 200 bytes of data/read. The data assumptions are shown in Table 10.24.

Table 10.22   Home area network data

Field	Instance	Description	Bytes
Date	20120910	Measurement date	4
Time	113254	Measurement time	3
Readings	–	Readings on the consumption, voltage, etc.	383
Hours_Usage	12	Total hours of usage	4
Tariff	B	Tariff category	2
Appliance_ID	24	Smart appliance unique ID	4
TOTAL			400

Table 10.23   Smart water metering

Field	Instance	Description	Bytes
Date	20120910	Measurement date	4
Time	113254	Measurement time	3
Readings	–	Readings on the consumption, voltage, etc.	249
Meter_ID	24	Meter unique ID	4
TOTAL			260

Table 10.24   Water sensing

Field	Instance	Description	Bytes
Date	20120910	Measurement date	4
Date	20120910	Measurement date	4
Time	113254	Measurement time	3
Dissolved oxygen	50	Milligrams (mg) of oxygen	4
pH	6	Water pH	4
Conductivity	0.005	Siemens per meter	8
Temperature	10	Water temperature (°C)	4
Chlorophyll	1	Generic fluorescence units or μg/L of chlorophyll	4
Turbidity	0.02	Nephelometric turbidity unit	8
Ammonium	2.0	Total ammonia (mg/L)	8
Sensor_ID	24	Meter unique ID	4
Longitude	114.874565	Degree	31
Latitude	22.784498	Degree	31
TOTAL			112

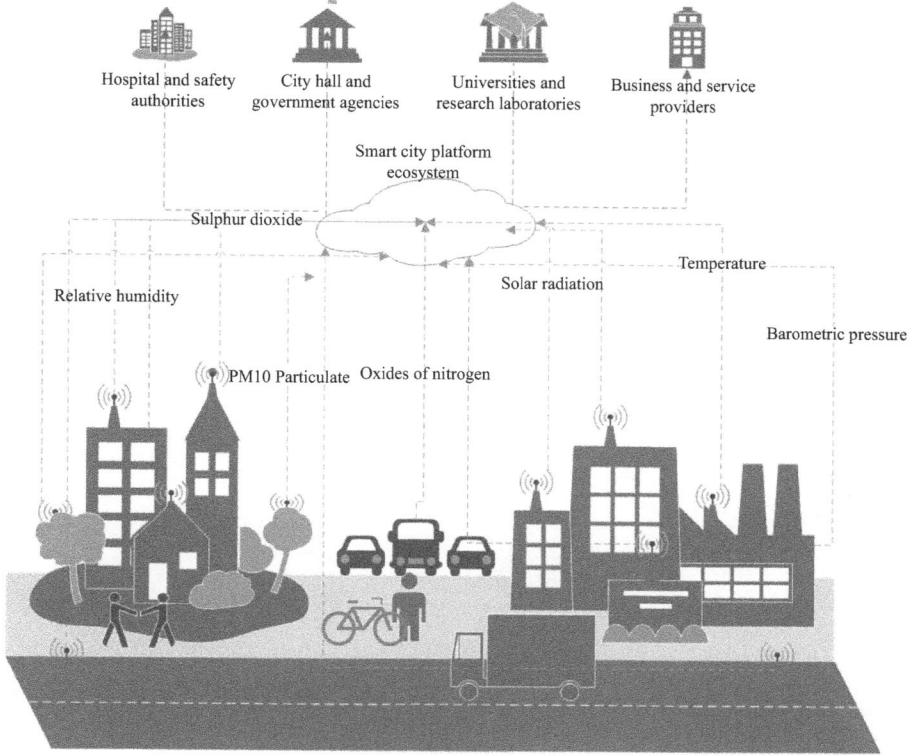

Figure 10.3   Waste and environment data

10.3.1.3   Waste and environment

In this section, we describe the following waste and environment datasets: smart bins, smart waste tracking and smart environment tracking. Figure 10.3 illustrates these data sets on the urban environment.

SMART BINS: Smart bins are litter bins embedded with sensors that will measure the level of rubbish inside the bins. When the bins are full or about to be, waste company can route the truck to these specific bins. We assume that Greater London's litter bins are embedded with sensors. Each smart bin sensor is read once/hour and generates 27 bytes per message. We consider 33,000 smart bins as there are approximately 1,000 of bins in each London Borough. The data assumptions are shown in Table 10.25 and illustrated in Figure 10.4.

ENVIRONMENT SENSING: We have assumed that there are 5 sensors of each one of the following 14 environmental sensors in all London Boroughs: barometric pressure (mBar), nitric oxide (μg/m^3), nitrogen dioxide (μg/m^3), oxides of nitrogen (μg/m^3 as NO^2), ozone (μg/m^3), PM10 particulate (by FDMS) (μg/m^3), PM2.5 particulate (by FDMS) (μg/m^3), rainfall (mm), relative humidity (%),

Table 10.25 Smart bins data

Field	Instance	Description	Bytes
Date	20120910	Measurement date	4
Time	113254	Measurement time	3
Reading	–	Volume of trash	8
Bin_ID	24	Bin unique ID	4
Reading	–	Volume of trash	8
TOTAL			27

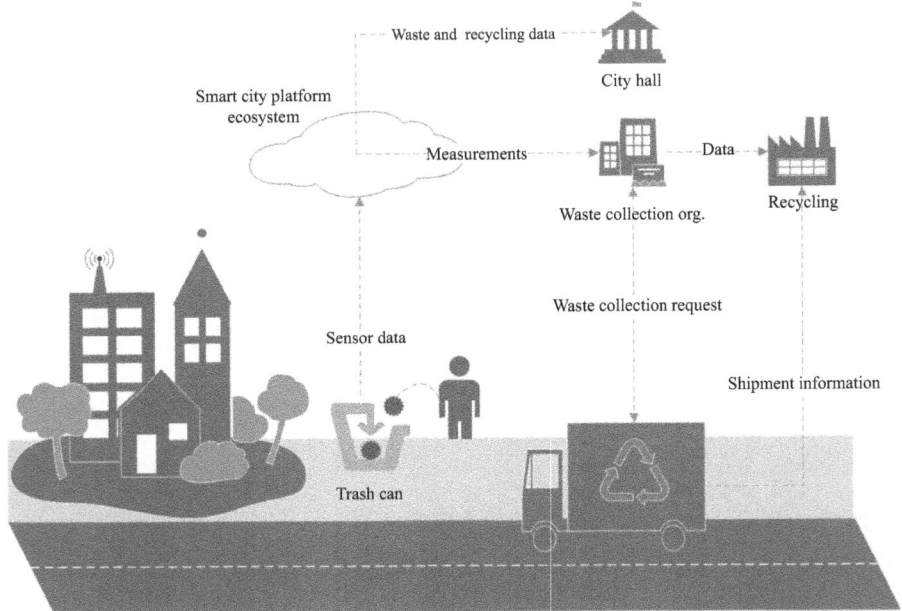

Figure 10.4 Smart bins data

solar radiation (W/m^2), sulphur dioxide (μg/m^3), temperature (°C), wind direction (°N) and wind speed (m/s). We consider 42 bytes of data per sensor reading and one reading per minute. The data assumptions are shown in Table 10.26.

SMART STREET LIGHTS: We have assumed that there are 13,800 km and the minimum distance between street lights is 10 m. We consider that 10% of these street lights are smartly controlled and generate 180 bytes of data 10 times/day. The data assumptions are shown in Table 10.27.

SMART WASTE TRACKING: We have assumed that toxic and hazardous waste have one RFID sensor to track its location and condition, and each message would typically generate 73 bytes. Each waste set is be tracked at every hour.

Table 10.26   Smart bins data

Field	Instance	Description	Bytes
Date	20120910	Measurement date	4
Time	113254	Measurement time	3
Sensor_ID	24	Sensor unique ID	4
Readings	–	$\mu g/m^3$, W/m^2, m/s, etc.	31
TOTAL			42

Table 10.27   Smart bins data

Field	Instance	Description	Bytes
Date	20120910	Measurement date	4
Time	113254	Measurement time	3
ID	23	Sensor unique ID	4
Readings	-	E.g. status of the lights	169
TOTAL			180

For estimation purposes, we consider that every day 1,000 toxic and hazardous waste set are tracked in Greater London. The data assumptions are shown in Table 10.28 and illustrated in Figure 10.5.

10.3.2   Simulation of CDIs

In the case study '*London Data Infrastructure*', the CSF **Increase responsiveness of systems** and **Acceptable Quality of Service** were evaluated negatively. This is due to the fact that some scalability issues were not identified in the business models, and such issues may cause the system to be non-responsive and reduce the quality of services of the platform. Even though the CDI of scalability is not marked as critical in the business model, it is an essential CDI in the case of London Data Infrastructure. After examining the CSFs and we have considered that scalability should be included as a CDI in the technical domain.

In order to assess the suitability and scalability of a data infrastructure to-be, a series of software simulation experiments can be performed. We simulated and evaluated the performance of an architecture for city data management using the component-based software simulation method. To simulate the architecture, we used the Palladio Component-based software simulator (PCM) [423] as it provides means to simulate component performance models and output prediction metrics.

PCM features a meta-model based on the Eclipse Modelling Framework (EMF) and provides specialized modelling languages for component developers and software architects. The component performance models used by PCM are parameterized for different usage profiles, hardware resources and required services as described above. After different component performance models have been composed into

Table 10.28   Smart bins data

Field	Instance	Description	Bytes
Date	20120910	Measurement date	4
Time	113254	Measurement time	3
Object_ID	24	Object unique ID	4
Longitude	114.874565	Degree	31
Latitude	22.784498	Degree	31
TOTAL			73

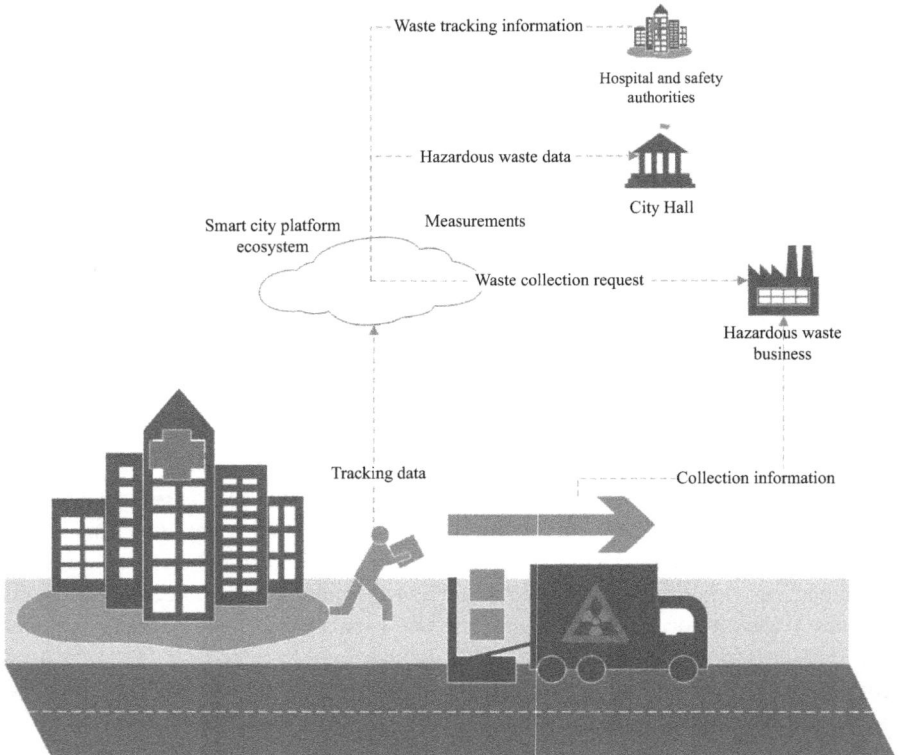

Figure 10.5   Smart waste tracking data

an architectural model, they are transformed into a discrete event simulation (SimuCOM) to derive the desired performance metrics. The PCM has been applied in various industrial projects (see [423]). The parameterized component performance models (i.e. behavioural specification with resource demands) are retrieved from a repository and composed of a software architect depending on the desired application architecture. Additionally, the system-level workload of the application as well as the hardware resource environment has to be described to complete the annotated architectural model. The component-based method effort for creating parameterized models is very high; however, it also promises that the effort can pay off after the model has been reused only a few times.

In component-based software engineering (CBSE), the development of a software system is typically distributed over multiple independent roles. Each role takes different responsibilities and contributes to the overall software system. In the context of the PCM, we distinguish four developer roles that produce artefacts of a software system [209]. The experiments were conducted in a MacAir station running an Intel Core i5-2557M CPU @ 1.7 GHz 1.70 GHz and with 4 GB memory. The analysed systems were simplified versions of realistic component-based software for city data management as we have just assumed the required software functionalities and the required resource demand for each component. To illustrate the capabilities of PCM in simulating software architectures, we executed an analytical Markov chain solver for different BRS usage scenarios and design alternatives. Figure 6.14 illustrates a simplistic view of one of the software component diagrams of the London Data Infrastructure designed in PCM.

For illustration purposes, we present here the results obtained by having 1,000 user inter-arrival rate using different types of data serialization: TRIX, TRIG and RDF, using a cache and index database. Results: User inter-arrival rate: 1,000 users. The proposed deployment model and architecture organization is able to scale to serve 90% of the users (both data providers and consumers) within 3 min. Although we used the mentioned approximations of the parametric dependencies, the measured and the simulation results still match quite well. Moreover, we are able to see that the design alternative with using TRIX serialization was much faster than TRIG and RDF serialization. Figure 10.6 illustrates the response time for using TRIG serialization (in milliseconds) and cache database. Hence, to maintain the scalability of the platform, a cache database and TRIG serialization should be used.

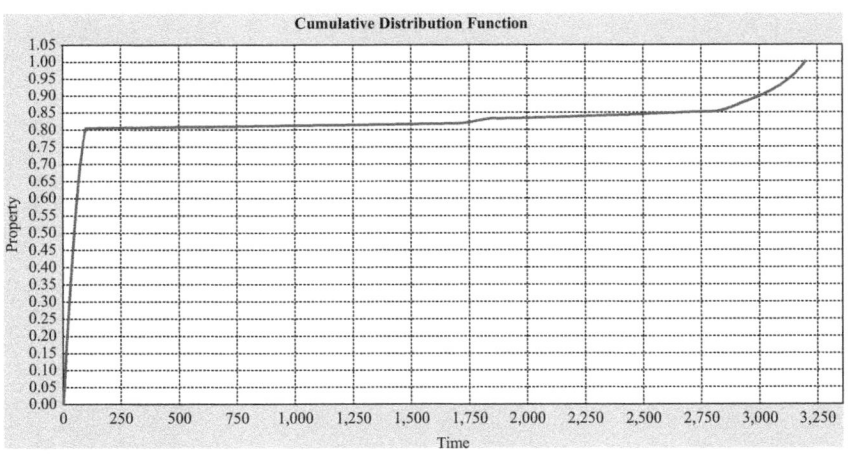

Figure 10.6 Response time for using TRIG serialization (in milliseconds) and cache database

Chapter 11

Conclusion

This book concentrates on the definition of data infrastructures and their business models and components. Our method employs a business model-driven approach to support the elicitation and modelling of requirements and data strategies, and a closed-loop supply chain model to serve as a reference architecture model for data infrastructures. By using critical design issues and factors, the positive and negative contributions that may occur among the requirements and specific design needs can be easily identified, as well as the final contributions of the data infrastructure to the realisation of smart cities. Our framework facilitates the requirements of elicitation process from business models analysis, and the detection of requirements mismatches across the five domains of the business models. It offers templates for requirements balancing and refinement which can be used to determine the trade-offs to be made during the design of such large interconnected systems.

The dynamic business models' approach enables decision-makers to evaluate the evolution of the business models and how external factors may impact the several stages of the development process of digital strategies of cities. The closed-loop supply chain model can give government and their value network of partners the ability to better collaborate on the basis of a common and accurate reference architecture. As such, cities are equipped with methodologies that facilitate the design and delivery of sustainable, agile, evolvable and cost-effective data management solutions for smart cities.

Data infrastructures have a profound and positive influence on the smart cities vision set by cities around the world. The moral purpose of data infrastructure is to contribute for creating a prosperous and sustainable economy. Governments and technology corporations often forget this basic truth. When governments disregard the efforts made by local initiatives within their cities, they penalize productive clusters of stakeholders of smart cities who are working towards the same goal. Such governments and clusters of bottom-up approaches are fated to financial loss, services bias and unfairness, information asymmetry, standards and regulations nonconformity, and hindering the city's ability to leverage its collective knowledge and serve all of society.

Our framework provides to governments and supporting and contributing partners with the clarity they need to change this scenario. It, however, does not intend to describe a one-size-fits-all model for data infrastructures in smart cities. Rather, the focus is on the enabling processes by which innovative use of technology and data coupled with governance strategies, contextualised and inclusive engaging urban services, a strong value network of partners can help deliver the various visions of data strategies for cities in more efficient, aligned and effective ways.

Efforts to find shared value in data infrastructures have the potential not only to foster economic and support the delivery of smart cities outcomes development but also to change the way government and independent clusters of initiatives related to each other.

We understand that it will require dramatically different thinking in governments. We are convinced, however, that data infrastructure will become increasingly important to the success of the smart cities strategies. Governments neither specialise in all the capabilities required to deliver data infrastructures and solve all city data integration problems nor do they have the resources to solve them all. Each member of the data infrastructure value network can identify the particular set of problems that it is best equipped to help resolve and from which it can gain the greatest competitive and economic benefit.

Supporting smart cities initiatives by creating a shared data infrastructure will lead to self-sustaining and innovative solutions that will create the cities of the future. Data infrastructures can provide much functionality that transcend space (and time), break down the barriers to information access and enhance communication and collaboration among government, businesses and citizens. As a result, it enables stakeholders of city data to have access to information that will enable them to innovate, to work better, to commute more efficiently in between places, enable governments to get insights on the urban services being provided anywhere and anytime they want.

When cities orchestrate their data infrastructure in a value network, the stakeholders of their many dispersed city data initiatives apply its vast resources, expertise to problems that it understands and in which it has a stake, and cities can take advantage of unprecedented insights into how the city and its infrastructure functions and be ready to overcome social challenges.

We encourage the smart cities community to adopt the ideas presented in this book when building their data infrastructure and to incorporate our frameworks into their analysis and design process. Finally, we encourage the academic community to concentrate more research effort into the interesting and important fields of business models, human–computer interaction, ethics in artificial intelligence and algorithm fairness, urban data management and digital services platforms.

References

[1]   Caragliu A, Del Bo C, Nijkamp P. Smart Cities in Europe. Free University Amsterdam, Faculty of Economics, Business Administration and Econometrics; 2009.

[2]   Dirks S, Gurdgiev C, Keeling M. Smarter Cities for Smarter Growth: How Cities Can Optimize Their Systems for the Talent-Based Economy. IBM Global Business Services; 2010.

[3]   Hollands RG. Will the real smart city please stand up? City. 2008;12(3): 303–320.

[4]   Giffinger R, Fertner C, Kramar H, et al. Smart Cities: Ranking of European Medium-Sized Cities. Vienna University of Technology; 2007.

[5]   Mitchell W. City of Bits: Space, Place, and the Infobahn. 1st ed. Cambridge: MIT Press; 1996.

[6]   Negroponte N. Being Digital. New York: Vintage Books; 1996.

[7]   Mitchell W. E-topia. Cambridge: MIT Press; 2000.

[8]   Martin J. The Wired Society. New York: Prentice Hall; 1978.

[9]   Castells M. The Rise of the Network Society. Oxford: Blackwell; 2000.

[10] Foth M. Handbook of Research on Urban Informatics: The Practice and Promise of the Real-Time City. Information Science Reference. Hershey, PA; 2009.

[11] Satterthwaite D. The Earthscan Reader in Sustainable Cities. London: Earthscan Publications Ltd.; 2001.

[12] Graham S, Marvin S. Splintering Urbanism, Networked Infrastructures, Technological Mobilities and the Urban Condition. London: Taylor and Francis; 2001.

[13] Rutherford J. Rethinking the relational socio-technical materialities of cities and ICTs. Journal of Urban Technology. 2011;18(1):21–33.

[14] Jacobs J. The Economy of Cities. New York: Vintage Books; 1969.

[15] Bairoch P. Cities and Economic Development. Chicago: University of Chicago Press; 1988.

[16] Simmie J. Innovative Cities. London: Spon; 2001.

[17] Modelski G. World Cities, −3000 to 2000. Washington, DC: FAROS 2000; 2003.

[18] Glaeser E. Triumph of the City. New York: Penguin Press; 2011.

[19] Bose RK. Energy Efficient Cities: Assessment Tools and Benchmarking Practices. Washington, DC: World Bank; 2009.

[20] Demirbas A. Global renewable energy projections. Energy Sources, Part B: Economics, Planning, and Policy. 2009;4(2):212–224.

[21]   Nye D. American Technological Sublime. Cambridge, MA: MIT Press; 1996.

[22]   Gann DM, Dodgson M, Bhardwaj D. Physical-digital integration in city infrastructure. IBM Journal of Research and Development 2011;55(1 and 2):90–99.

[23]   Van Den Besselaar P, Beckers D. The Life and Death of the Great Amsterdam Digital City. In: Van Den Besselaar P, Koizumi S, editors. Digital Cities III. Information Technologies for Social Capital: Cross-cultural Perspectives. Berlin/Heidelberg: Springer; 2005.

[24]   Lindskog H. Smart Communities Initiatives. In *Proceedings of the 3rd ISOneWorld Conference*, Las Vegas, NV, April 14–16; 2004. Available at http://www.heldag.com/articles/Smart%20communities%20april%202004. pdf

[25]   Mitchell W. Intelligent Cities. Universitat Oberta de Catalunya; 2007.

[26]   Marceau J. Innovation in the city and innovative cities. Innovation: Management, Policy and Practice. 2008;10(2/3):136–145.

[27]   Yigitcanlar T. Planning for smart urban ecosystems: Information technology applications for capacity building in environmental decision making. Theoretical and Empirical Researches in Urban Management. 2009;4(3(12)): 5–21.

[28]   Giraud PN, Benard M, Lefevre B, *et al.* Energy and Urban Innovation. World Energy Council; 2010. Available from: http://www.worldenergy.org/ wpcontent/uploads/2012/10/PUB_Energy_and_urban_innovation_2010_ WEC.pdf. Accessed September 16, 2014.

[29]   Gilding P. The Great Disruption: How the Climate Crisis Will Transform the Global Economy. London: Bloomsbury Publishing Plc; 2011.

[30]   Drucker PF. Post Capitalist Society. New York, NY; 1993.

[31]   Toppeta D. The Smart City Vision: How Innovation and ICT Can Build Smart, Liveable, Sustainable Cities. Think! Report; 2010.

[32]   Greenburg E. Codifying New Urbanism: How to Reform Municipal Land Development Regulations. American Planning Association PAS; 2004.

[33]   Chandler AD. Scale and Scope: The Dynamics of Industrial Capitalism. Cambridge: Harvard University Press; 1990.

[34]   Greif. Comparative and Historical Institutional Analysis: A Game Theoretical Perspective. Cambridge University Press; 2003.

[35]   North DC, Weingast B. Constitutions and commitment: The evolution of institutions governing public choice in seventeenth century England. Journal of Economic History. 1989;49(4):803–832.

[36]   Shleifer A, Vishny R. The Grabbing Hand: Government Pathologies and Their Cures. Cambridge, MA: Harvard University Press; 1998.

[37]   Baumol W. The Free-Market Innovation Machine: Analyzing the Growth Miracle of Capitalism. Princeton, NJ: Princeton University Press; 2002.

[38]   Mowery DC, Rosenberg N. Technology and the Pursuit of Economic Growth. Cambridge: Cambridge University Press; 1989.

[39]   Karvonen E, editor. Informational Societies. Understanding the Third Industrial Revolution. Tampere University Press; 2001.

[40] Freeman C, Soete L. Work for All or Mass Unemployment. London: Pinter Publishers; 1994.

[41] Halpern D. Social Capital. Cambridge: Polity Press; 2005.

[42] Putnam RD. Bowling alone: America's declining social capital. Journal of Democracy. 1993;6:65–78.

[43] Graham S. The Cybercities Reader. London: Routledge; 2004.

[44] Florida R. The Rise of the Creative Class and How It's Transforming Work, Leisure, Community and Everyday Life. New York: Basic Books; 2002.

[45] Lundvall BA. Product Innovation and User–Producer Interaction. Aalborg: Aalborg University; 1985.

[46] Lundvall BA, Johnson B. The learning economy. Journal of Industry Studies. 1994;1(2):23–42.

[47] Santinha G, Castro EA. Creating more intelligent cities: The role of ICT in promoting territorial governance. Journal of Urban Technology. 2010; 17(2).

[48] Morgan K. The learning region: Institutions, innovation and regional renewal. Regional Studies. 1997;31(5):491–503.

[49] Marshall D, Toffel W. Framing the elusive concept of sustainability: A sustainability hierarchy. Environmental Science and Technology. 2005;39(3): 673–682.

[50] Hall RE. The vision of a smart city. In: 2nd International Life Extension Technology Workshop, Paris, France; 2000.

[51] Adams WM, Jeanrenaud SJ. Transition to Sustainability: Towards a Humane and Diverse world. Gland: IUCN; 2008.

[52] Komninos N. The Architecture of Intelligent Cities: Integrating Human, Collective and Artificial Intelligence to Enhance Knowledge and Innovation. In: 2nd IET International Conference on Intelligent Environments IE 06, London, UK, vol. 1; 2006. p. 13–20.

[53] Qi L, Shaofu L. Research on digital city framework architecture. In: IEEE International Conferences on Info-Tech and Info-Net, Beijing, China; 2001. p. 30–36.

[54] Yovanof G, Hazapis G. An Architectural Framework and Enabling Wireless Technologies for Digital Cities and Intelligent Urban Environments. Wireless Personal Communications. 2009;49(3):445–463.

[55] Ishida T, Aurigi A, Yasuoka M. World Digital Cities: Beyond Heterogeneity. Springer-Verlag; 2005.

[56] Ishida T, Isbister K. Digital Cities, Technologies, Experiences, and Future Perspectives. Lecture Notes in Computer Science. vol. 1765. Springer-Verlag; 2000.

[57] Kuk G, Janssen M. The business models and information architectures of smart cities. Journal of Urban Technology. 2011;18(2):39–52.

[58] Sairamesh J, Lee A, Anania L. Information cities. Communications of the ACM. 2004;47(2):28–31.

[59] Sproull L, Patterson JF. Making information cities livable. Communications of the ACM. 2004;47(2):33–37.

[60]   Widmayer P. Building digital metropolis: Chicago's future networks. IT Professional. 1999;1(4):40–46.

[61]   Evans D S, Hagiu A, Schmalensee R. Invisible Engines: How Software Platforms Drive Innovation and Transform Industries. Cambridge, Mass.: MIT Press; 2006.

[62]   Thomas R. Sustainable Urban Design – An Environmental Approach. New York: Spon Press; 2003.

[63]   Tocci G. Governance, Cities and New ICTs: E-Democracy in European Urban Contexts. In: 17th NISPAcee Annual Conference; 2009.

[64]   Pina V, Torres L, 2010. E-governance developments in European Union cities: Reshaping government's relationship with citizens. International Journal of Policy. 2010;19(2):277–302.

[65]   Ishida T. Digital city Kyoto. Communications of the ACM. 2002;45: 76–81.

[66]   Schuler D. The Seattle Community Network, Anomaly or Replicable Model? In: Van Den Besselaar P, Koizumi S, editors. Proceedings of the Third International Conference on Information Technologies for Social Capital: Cross-Cultural Perspectives. Berlin/Heidelberg/New York: Springer-Verlag; 2005.

[67]   Peng D, Lin DH, Sheng HY. In: Van Den Besselaar P, Koizumi S, editors. Digital City Shanghai. Lecture Notes in Computer Science. Berlin/Heidelberg/New York: Springer-Verlag; 2005.

[68]   Linturi R, Simula T. Virtual Helsinki: Enabling the Citizen, Linking the Physical and Virtual. In: Van Den Besselaar P, Koizumi S, editors. Digital Cities III. Information Technologies for Social Capital: Cross-cultural Perspectives. Berlin/Heidelberg/New York: Springer-Verlag; 2005.

[69]   Boschma R. Competitiveness of regions from an evolutionary perspective. Regional Studies. 2004;38:1001–1014.

[70]   Middleton CA, Bryne A. An exploration of user-generated wireless broadband infrastructures in digital cities. Telematics and Informatics. 2011;28(3): 163–175.

[71]   Ergazakis E, Ergazakis K, Askounis D, *et al.* Digital cities: Towards an integrated decision support methodology. Telematics and Informatics. 2011;28(3):148–162.

[72]   Lin Sf, Li Q. Research on the architecture of digital city application service platform. In: International Conferences on Info-tech and Info-net. vol. 1; 2001. p. 37–43.

[73]   Gang W, Zhiping Z, Xiaodong Z, *et al.* Research on Integrated Digital City Model and Its Application. In: The Second International Conference on Computer Engineering and Technology (ICCET), Chennai, India, vol. 4; 2010. p. V4–466–V4–470.

[74]   Jun Z, Jiatao J, Yi Y, *et al.* Ontology-Based Integrated Information Platform for Digital City. In: 4th International Conference on Wireless Communications, Networking and Mobile Computing, WiCOM'08, Dubai, UAE; 2008. p. 1–4.

[75] Jiaxing L, Yinsheng L, Xiaohua L. Service-Oriented and Collaborative Portal for Digital Cities. In: 11th International Conference on Computer Supported Cooperative Work in Design, CSCWD 2007, Melbourne, Australia; 2007. p. 531–537.

[76] Zhu D, Li Y, Shi J, *et al.* A service-oriented city portal framework and collaborative development platform. Information Sciences. 2009;179(15): 2606–2617.

[77] Maoqing F, Rugang W. Research on Framework of Integration Platform for Digital Urban Planning based- on Service-oriented Architecture. In: The 2nd IEEE International Conference on Information Management and Engineering (ICIME); 2010. p. 207–209.

[78] Dingju Z, Jianping F. Application of Parallel Computing in Digital City. In: 10th IEEE International Conference on High Performance Computing and Communications, 2008 (HPCC '08); 2008. p. 845–848.

[79] Komninos N, Schaffers H, Pallot M. Developing a Policy Roadmap for Smart Cities and the Future Internet. In: Cunningham PC, editor. eChallenges e-2011 Conference Proceedings. Florence: IIMC International Information Management Corporation; 2011.

[80] Graham S. Bridging urban digital divides? Urban polarisation and information and communications technologies (ICTs). Urban Studies. 2002;39(1): 33–56.

[81] Sheller M. Mobile publics: Beyond the network perspective. Environment and Planning D: Society and Space. 2004;22:39–52.

[82] Weiser M. Hot topics-ubiquitous computing. Computer. 1993;26:71–72.

[83] Cook DJ, Das SK. Smart Environments: Technology, Protocols and Applications. Hoboken, NJ: John Wiley and Sons, Inc.; 2005.

[84] Johanson B, Fox A, Winograd T. The Interactive Workspaces Project: Experiences with Ubiquitous Computing Rooms. IEEE Pervasive Computing. 2002;1(2):67–74.

[85] Satyanarayanan M. Pervasive Computing: Vision and Challenges. IEEE Personal Communications. 2001;8(4):10–17.

[86] Bellavista P, Corradi A. The Handbook of Mobile Middleware. Boston, MA: CRC Press; 2007.

[87] Lee S, Yigitcanlar T, Hoon H, *et al.* Ubiquitous urban infrastructure: Infrastructure planning and development in Korea. Innovation: Management, Policy and Practice. 2008;10(3):282–292.

[88] Lee S. Ubiquitous city and urban planning. Illuminating and Electrical Installation. 2005;19(4):29–42.

[89] Choi J. The city is connections: Seoul as an urban network. Multimedia Systems. 2010;16:75–84.

[90] Song LL. IT389 Policy leading to U-Korea. Ministry of Information and Communication, Republic of Korea; 2006.

[91] Kwon O, Kim J. A Methodology of Identifying Ubiquitous Smart Services for U-city Development. In: Indulska J, Ma J, Yang LT, Ungerer T, Cao J, editors. Ubiquitous Intelligence and Computing. Berlin: Springer; 2007.

[92]   Jang M, Suh ST. U-City: New Trends of Urban Planning in Korea Based on Pervasive and Ubiquitous Geotechnology and Geoinformation. In: Taniar D, Gervasi O, Murgante B, *et al.*, editors. Computational Science and Its Applications, ICCSA 2010. Lecture Notes in Computer Science. Springer; 2010.

[93]   Moss M, Townsend A. How Telecommunications Systems Are Transforming Urban Spaces. In: Wheeler J, Aoyama Y, Warf B, editors. Cities in the Telecommunications Age: The Fracturing of Geographies. London: Routledge; 2000.

[94]   Ha W. A study on action plan and strategies for building U-Korea: Vision, issue, task, and system. Telecommunications Review. 2003;13(1): 4–15.

[95]   Jung C, Shin Y, Shin D, *et al.* Study of Security Reference Model of u-city Integrated Operating Center. World Scientific and Engineering Academy and Society (WSEAS); 2009.

[96]   Anthopoulos L, Fitsilis P. From Digital to Ubiquitous Cities: Defining a Common Architecture for Urban Development. In: Sixth International Conference on Intelligent Environments (IE), Kuala Lumpur, Malaysia; 2010. p. 301–306.

[97]   Yong Woo L, Seungwoo R. U-city portal for smart ubiquitous middleware. In: 12th International Conference on Advanced Communication Technology (ICACT), Gangwon-Do, South Korea, vol. 1; 2010. p. 609–613.

[98]   Jung HS, Jeong CS, Lee YW, *et al.* An intelligent ubiquitous middleware for U-city: SmartUM. Journal of Information Science and Engineering. 2009;25(2):375–388.

[99]   Shin D. Ubiquitous city: Urban technologies, urban infrastructure and urban informatics. Journal of Information Science. 2009;35:515–526.

[100] Sassen S. The Global City: New York, London, Tokyo. Princeton: Princeton University Press; 2001.

[101] Eger J. Cyberspace and Cyberplace: Building the smart communities of tomorrow. San Diego Union-Tribune, Insight. 1997;2.

[102] Dutton WH, Kraemer KL, Blumler JG. Wired cities: Shaping the future of communications. Boston, MA: Macmillan Publishing Co., Inc.; 1987.

[103] Webster F. Information and the Idea of an Information Society. Theories of the Information Society. London: Routledge. 1995;p. 6–51.

[104] Aurigi A. Making the digital City: The Early Shaping of Urban Internet Space. Aldershot: Ashgate; 2005.

[105] Tanabe M, van den Besselaar P, Ishida T. Digital Cities II: Computational and Sociological Approaches. Berlin: Springer; 2002.

[106] Koizumi S, van den Besselaar P. Digital Cities III: Information Technologies for Social Capital – Cross-Cultural Perspectives. Berlin/Heidelberg/New York: Springer-Verlag; 2005.

[107] Malek JA. Informative Global Community Development Index of Informative Smart City. In: Eighth WSEAS International Conference on Education and Educational Technology. Genova, Italy; 2009.

[108]   Komninos N. Intelligent Cities and Globalization of Innovation Networks. London and New York: Routledge; 2008.

[109]   De Moor BKea K. User Involvement in Living Labs Research: Experiences from an Interdisciplinary Study on Future Mobile Applications. In: Third International Seville Seminar on Future-Oriented Technology Analysis: Impacts and implications for policy and decision-making. Seville; 2008.

[110]   Bergvall KB, St Pst A. Living lab: An open and citizen-centric approach for innovation. International Journal of Innovation and Regional Development. 2009;1(4):356–370.

[111]   Gloor P. Swarm Creativity: Competitive Advantage Through Collaborative Innovation Networks. Oxford: University Press; 2006.

[112]   Komninos N, Sefertzi E. Intelligent Cities: RandD Offshoring, Web 2.0 Product Development and Globalization of Innovation Systems; 2009.

[113]   Moulaert F, Sekia F. Territorial Innovation Models: A Critical Survey. Regional Studies. 2003;31(5):491–503.

[114]   Kanter MR, Litow S. Informed and Interconnected: A Manifesto for Smarter Cities. Harvard Business School Working Paper; 2009.

[115]   Atkinson RD, Castro D. Digital Quality of Life: Understanding the Personal and Social Benefits of the Information Technology Revolution. Social Science Research Network; 2008.

[116]   Bellissent J. Getting clever about smart cities: New opportunities require new business models. Forrester for Vendor Strategy Professionals; 2010.

[117]   Shapiro JM. Smart Cities: Explaining the Relationship between City Growth and Human Capital; 2003.

[118]   Blewitt J. Understanding Sustainable Development. London: Earthscan; 2008.

[119]   Allwinkle S, Cruickshank P. Creating smarter cities: An overview. Journal of Urban Technology. 2011;18(2):1–16.

[120]   Lee G, Kwak YH. An open government maturity model for social media-based public engagement. Government Information Quarterly. 2012;29(4): 492–503.

[121]   Washburn D, Sindhu U, Balaouras S, *et al*. Helping CIOs Understand Smart City Initiatives: Defining the Smart City, Its Drivers, and the Role of the CIO. Cambridge, MA: Forrester Research, Inc; 2010.

[122]   Dameri RP. Searching for Smart City definition: A comprehensive proposal. International Journal of Computers & Technology. 2013;11(5).

[123]   Su LJ K, Fu H. Smart City and the Applications. In: IEEE International Conference on Electronics, Communications and Control (ICECC); 2011. p. 1028–1031.

[124]   Falconer G, Mitchell S. Smart City Framework: A Systematic Process for Enabling Smart+Connected Communities. CISCO; 2012. Available from: https://www.cisco.com/web/about/ac79/docs/ps/motm/Smart-City-Framework.pdf.

[125]   Hall P, Pfeiffer U. Urban Future 21: A Global Agenda for Twenty-First Century Cities. London: Spon; 2000.

[126]   Cosgrove M, Harthoorn Wea. Smart Cities Series: Introducing the IBM City Operations and Management Solutions; 2011. Available from: http://www.redbooks.ibm.com/redpapers/pdfs/redp4734.pdf.

[127]   Haque U. Surely There's a Smarter Approach to Smart Cities? 2012. Available from: http://www.wired.co.uk/news/archive/2012-04/17/potential-of-smartercities-beyond-ibm-and-cisco.

[128]   Partridge H. Developing a human perspective to the digital divide in the smart city. In: Biennial Conference of Australian Library and information Association; 2004.

[129]   Rios P. Creating "The Smart City." University of Detroit; 2008. Available from: https://archive.udmercy.edu/bitstream/handle/10429/393/2008_rios_smart.pdf?sequence=1.

[130]   IBM; 2011.

[131]   Giffinger R, Gudrun H. Smart cities ranking: An effective instrument for the positioning of cities? ACE: Architecture, City and Environment. 2010;4(12):7–25.

[132]   Naphade M, Banavar G, Harrison C, *et al*. Smarter cities and their innovation challenges. Computer. 2011;44(6):32–39.

[133]   Klein C, Kaefer G. From Smart Homes to Smart Cities: Opportunities and Challenges from an Industrial Perspective. Berlin, Heidelberg: Springer-Verlag; 2008.

[134]   Francesco Bifulco CCA Marco Tregua, D'Auria A. ICT and sustainability in smart cities management. Journal of Public Sector Management. 2016;29(2):132–147.

[135]   D'Onofrio S, Portmann E. Cognitive computing in Smart Cities. Informatik Spektrum. 2017;40(1).

[136]   Harrison C, Eckman B, Hamilton R, *et al*. Foundations for Smarter Cities. IBM Journal of Research and Development. 2010;54(4).

[137]   Pallot TBSBSH M, Komninos N. Future Internet and Living Lab Research Domain Landscapes: Filling the Gap between Technology Push and Application Pull in the Context of Smart Cities. In: Cunningham P, Cunningham M, editors. eChallenges e-2011 Conference Proceedings. IIMC International Information Management Corporation; 2011.

[138]   Rotmans J. In: Ruth M, editor. A Complex systems approach for Sustainable cities. Cheltenham: Edward Elgar Publishing Limited; 2006.

[139]   Kotov V. Systems of Systems as Communicating Structures. Technical Report HPL-97-124. Hewlett Packard Computer Systems Laboratory Paper; 1997.

[140]   Lukasik SJ. Systems, systems of systems, and the education of engineers. Artificial Intelligence for Engineering Design, Analysis and Manufacturing 1998;12(1):55–60.

[141]   Sage AP, Cuppan CD. On the Systems Engineering and Management of Systems of Systems and Federations of Systems. Information Knowledge Systems Management 2001;2(4):325–345.

[142]   Selberg SA, Austin MA. Toward an Evolutionary System of Systems Architecture. Technical Report. Institute for Systems Research. INCOSE; 2008.

[143] K A. That "The Internet of Things" thing; 2009. ITU Internet Report 2005: The Internet of Things.

[144] Cash DW, Clark WC, Alcock F, *et al*. Knowledge Systems for Sustainable Development. In: Proceedings of the National Academy of Sciences of the USA. vol. 100 of Science and Technology for Sustainable Development Special Feature; 2003. p. 8086–8091.

[145] Lawton G. Machine-to-machine technology gears up for growth. Computer. 2004;37:12–15.

[146] Atzori L, Iera A, Morabito G. The Internet of Things: A survey. Computer Networks. 2010;54(15):2787–2805.

[147] Ngai EWT, Moon KKL, Riggins FJ, *et al*. RFID research: An academic literature review (1995–2005) and future research directions. International Journal of Production Economics. 2008;112(2):510–520.

[148] Hong D, Chiu DK, Shen VY, *et al*. Ubiquitous enterprise service adaptations based on contextual user behavior. Information Systems Frontiers. 2007;9(4):343–358.

[149] Strassner M, Schoch T. Today's impact of ubiquitous computing on business processes. In: First International Conference on Pervasive Computing. Springer; 2002. p. 62–74.

[150] Moore GE. Cramming more components onto integrated circuits. Electronics. 1965;38(8).

[151] Palattella MR, Dohler M, Grieco A, et al. Internet of things in the 5G era: Enablers, architecture, and business models. IEEE Journal on Selected Areas in Communications. 2016;34(3).

[152] Simsek M, Aijaz A, Dohler M, Sachs J, Fettweis, G. 5G-enabled tactile Internet. IEEE Journal on Selected Areas in Communications. 2016;34(3): 460–473.

[153] Maier CMRBP M, Van DP. The tactile Internet: Vision, recent progress, and open challenges. IEEE Communication Magazine. 2016;54(5):138–145.

[154] Shi W, Dustdar S. The promise of edge computing. Computer. 2016; 49: 78–81.

[155] Taleb T, Dutta S, Ksentini A, Iqbal A, Flinck H. Mobile edge computing potential in making cities smarter. IEEE Communications Magazine. 2017;55(3):38–43.

[156] Delaney OGMP D T, Ruzzelli AG. Evaluation of Energy-Efficiency in Lighting Systems using Sensor Networks. In: First ACM Workshop on Embedded Sensing Systems for Energy-Efficiency in Buildings. ACM; 2009.

[157] Man-Ching Y, Ling-Jyh C, King I. A Survey of Human Computation Systems. In: International Conference on Computational Science and Engineering. Vancouver, BC, Canada, vol. 4; 2009. p. 723–728.

[158] Krause A, Horvitz E, Kansal A, *et al*. Toward Community Sensing. In: International Conference on Information Processing in Sensor Networks, IPSN'08, St. Louis, MO; 2008. p. 481–492.

[159] Ratti C, Pulselli RM, Williams S, *et al*. Mobile landscapes: Using location data from cell phones for urban analysis. Environment and Planning B: Planning and Design. 2006;33(5):727–748.

[160]   Reades J, Calabrese F, Sevtsuk A, *et al*. Cellular Census: Explorations in Urban Data Collection. IEEE Pervasive Computing 2007;6(3):30–38.

[161]   Reades J, Calabrese F, Ratti C. Eigenplaces: Analysing cities using the space-time structure of the mobile phone network. Environment and Planning B: Planning and Design. 2009;36(5):824–836.

[162]   Gonzalez MC, Hidalgo CA, Barabasi AL. Understanding individual human mobility patterns. Nature. 2008;453(7196):779–782.

[163]   McNamara L, Mascolo C, Capra L. Media sharing based on colocation prediction in urban transport. ACM; 2008.

[164]   Roth C, Kang SM, Batty M, *et al*. Structure of urban movements: Polycentric activity and entangled hierarchical flows. PLoS ONE. 2011;6(1).

[165]   Hey T, Trefethen A. In: The Data Deluge: An e-Science Perspective. John Wiley and Sons, Ltd; 2003. p. 809–824.

[166]   Zikopoulos PC, Eaton C, deRoos D, *et al*. Understanding big data. Analytics for enterprise class hadoop and streaming data. McGraw Hill; 2012.

[167]   Manyika J, Chui M, Brown B, *et al*. Big data: The next frontier for innovation, competition, and productivity. McKinsey Global Institute; 2011.

[168]   Borkar V, Carey MJ, Li C. Inside "Big Data management": Ogres, onions, or parfaits? Berlin: ACM; 2012.

[169]   Hopkins EB B. Expand Your Digital Horizon with Big Data. Forrester; 2011.

[170]   Jackson MC. Systems Thinking: Creative Holism for Managers. Chichester: Wiley; 2003.

[171]   Wen Y, Yang X, Xu Y. Cloud-Computing-based Framework for Multi-camera Topology Inference in Smart City Sensing System. ACM; 2010.

[172]   Filipponi L, Vitaletti A, Landi G, *et al*. Smart City: An Event Driven Architecture for Monitoring Public Spaces with Heterogeneous Sensors. In: 2010 Fourth International Conference on Sensor Technologies and Applications (SENSORCOMM); 2010. p. 281–286.

[173]   Al-Hader M, Rodzi A, Sharif AR, *et al*. Smart City Components Architecture. In: CSSim'09 International Conference on Computational Intelligence, Modelling and Simulation, 2009, Washington, DC; 2009. p. 93–97.

[174]   Asimakopoulou E, Bessis N. Buildings and Crowds: Forming Smart Cities for More Effective Disaster Management. In: Fifth International Conference on Innovative Mobile and Internet Services in Ubiquitous Computing (IMIS), Seoul, Korea; 2011. p. 229–234.

[175]   Mostashari A, Arnold F, Maurer M, *et al*. Citizens as sensors: The cognitive city paradigm. In: 8th International Conference and Expo on Emerging Technologies for a Smarter World (CEWIT), Hauppauge, NY; 2011. p. 1–5.

[176]   Hernandez-Munoz JM, Bernat Vercher J, Munoz L, *et al*. In: John D, Alex G, Anastasius G, *et al*., editors. Smart Cities at the Forefront of the Future Internet. Berlin, Heidelberg: Springer-Verlag; 2011. p. 447–462.

[177]   Andreini F, Crisciani F, Cicconetti C, *et al*. A Scalable Architecture for Geo-localized Service Access in Smart Cities. In: Future Network and Mobile Summit (FutureNetw), Warsaw, Poland; 2011. p. 1–8.

[178] Ratti C, Townsend A. Harnessing Residents' Electronic Devices Will Yield Truly Smart Cities. Scientific American Magazine; 2011.

[179] Anthopoulos L, Fitsilis P. Using Classification and Roadmapping techniques for Smart City viability realization. Electronic Journal of e-Government. 2013;11(1):326–336.

[180] Sanchez L, Galache JA, Gutierrez V, *et al*. SmartSantander: The meeting point between Future Internet research and experimentation and the smart cities. In: Future Network and Mobile Summit (FutureNetw), Warsaw; 2011. p. 1–8.

[181] Perera C, Zaslavsky A, Christen P, *et al*. Sensing as a Service Model for Smart Cities Supported by Internet of Things. Transactions on Emerging Telecommunications Technologies. 2014;25(1):81–93.

[182] Kakarontzas G, Anthopoulos LG, Chatzakou D, Vakali A. A Conceptual Enterprise Architecture Framework for Smart Cities – A Survey Based Approach; 2014 11th International Conference on e-Business (ICE-B), Vienna, 2014, pp. 47–54.

[183] Heath T, Bizer C. Linked data: Evolving the web into a global data space. Synthesis Lectures on the Semantic Web: Theory and Technology. 2011;1(1): 1–136.

[184] Haklay M, Weber P. OpenStreetMap: User-generated street maps. IEEE Pervasive Computing. 2008;7(4):12–18.

[185] Valle ED, Celino I, Dell'Aglio D. The experience of realizing a semantic web urban computing application. Transactions in GIS. 2010;14(2):163–181.

[186] Celino I, Contessa S, Corubolo M, *et al*. 3. In: Cudré-Mauroux P, Heflin J, Sirin E, *et al*., editors. Linking Smart Cities Datasets with Human Computation The Case of UrbanMatch. vol. 7650 of Lecture Notes in Computer Science. Berlin/Heidelberg: Springer; 2012. p. 34–49.

[187] Auer S, Bizer C, Kobilarov G, *et al*. 52. In: Aberer K, Choi KS, Noy N, *et al*., editors. DBpedia: A Nucleus for a Web of Open Data. vol. 4825 of Lecture Notes in Computer Science. Berlin/Heidelberg: Springer ; 2007. p. 722–735.

[188] Della Valle E, Celino I, Dell'Aglio D, *et al*. Semantic traffic-aware routing using the LarKC platform. IEEE Internet Computing. 2011;15(6):15–23.

[189] Attwood A, Merabti M, Fergus P, *et al*. SCCIR: Smart Cities Critical Infrastructure Response Framework. In: Developments in E-systems Engineering (DeSE); 2011. p. 460–464.

[190] Le-Phuoc D, Nguyen-Mau HQ, Parreira JX, *et al*. A middleware framework for scalable management of linked streams. Web Semantics: Science, Services and Agents on the World Wide Web. 2012;16(0):42–51.

[191] Lopez V, Kotoulas S, Sbodio M, *et al*. 10. In: Cudré-Mauroux P, Heflin J, Sirin E, *et al*., editors. QuerioCity: A Linked Data Platform for Urban Information Management. vol. 7650 of Lecture Notes in Computer Science. Berlin/Heidelberg: Springer; 2012. p. 148–163.

[192] Janssen M, Charalabidis Y, Zuiderwijk A. Benefits, adoption barriers and myths of open data and open government. Information Systems Management. 2012;29(4):258–268.

[193]   Hey T, Trefethen AE. Cyberinfrastructure for e-Science. Science. 2005; 308(5723):817–821.

[194]   Groth P, Sheng J, Miles S, *et al*. An Architecture for Provenance Systems. ECS, University of Southampton; 2006.

[195]   Goble C. Position Statement: Musings on Provenance, Workflow and (Semantic Web) Annotation for Bioinformatics; 2002.

[196]   Lee HM J H. Toward a Framework for Smart Cities: A Comparison of Seoul, San Francisco and Amsterdam. In: Smart Green Cities Conference: Innovations for Smart Green Cities: What's Working, What's Not, Stanford, CA; 2012.

[197]   O'Riain S, Curry E, Harth A. XBRL and open data for global financial ecosystems: A linked data approach. International Journal of Accounting Information Systems. 2012;13(2):141–162.

[198]   McLaren R, Waters R. Governing location information in the UK. The Cartographic Journal. 2011;48(3):172–178.

[199]   EuropeanComission. Communication from the Commission to the European Parliament, the Council, the European Economic and Social Committee and the Committee of the Regions. Open Data. An Engine For Innovation, Growth and Transparent Governance; 2011.

[200]   Jin X, Yiming H, Guojie L, *et al*. Metadata distribution and consistency techniques for large-scale cluster file systems. IEEE Transactions on Parallel and Distributed Systems. 2011;22(5):803–816.

[201]   Schuurman N, Deshpande A, Allen DM. Data integration across borders: A case study of the Abbotsford-Sumas aquifer (British Columbia/Washington State). 1. JAWRA Journal of the American Water Resources Association. 2008;44(4):921–934.

[202]   Conradie P, Choenni S. Exploring Process Barriers to Release Public Sector Information in Local Government. In: Proceedings of the 6th International Conference on Theory and Practice of Electronic Governance ACM; 2012.

[203]   Tuunainen VK, Tuunanen T. IISIn – A Model for Analyzing ICT Intensive Service Innovations in n-Sided Markets. In: 44th Hawaii International Conference on System Sciences (HICSS); 2011. p. 1–10.

[204]   Shapiro C, Varian HR. Information Rules: A Strategic Guide to the Network Economy. Cambridge, MA: Harvard Business School Press; 1999.

[205]   Savaskan RC, Bhattacharya S, Wassenhove LNv. Closed-loop supply chain models with product remanufacturing. Management Science. 2004;50(2):239–252.

[206]   Ghazawneh A, Henfridsson O. Balancing platform control and external contribution in third-party development: The boundary resources model. Information Systems Journal. 2013;23(2):173–192.

[207]   Williamson O, de Meyer A. Ecosystem advantage: How to successfully harness the power of partners. California Management Review. 2012;55(1):24–46.

[208]   Economides N, Katsamakas E. Two-sided competition of proprietary vs. open source technology platforms and the implications for the software industry. Management Science. 2006;52(7):1057–1071.

[209] Brosch F, Koziolek H, Buhnova B, *et al*. Architecture-based reliability prediction with the palladio component model. IEEE Transactions on Software Engineering. 2012;38(6):1319–1339.

[210] Basole R, Karla J. On the evolution of mobile platform ecosystem structure and strategy. Business and Information Systems Engineering. 2011;3(5): 313–322.

[211] Wiener N. Cybernetics. New York, NY: John Wiley and Sons; 1948.

[212] McCullough M. Digital Ground: Architecture, Pervasive Computing, and Environmental Knowing. Cambridge MA: MIT Press; 2004.

[213] Dourish P. Where the Action Is: The Foundations of Embodied Interaction,. Cambridge MA: MIT Press; 2004.

[214] Nam T, Pardo TA. Conceptualizing Smart City with Dimensions of Technology, People, and Institutions. New York: ACM; 2011.

[215] Patni H, Henson C, Sheth A. Linked Sensor Data. In: International Symposium on Collaborative Technologies and Systems (CTS), Chicago, IL; 2010. p. 362–370.

[216] Jesús BV, Santiago PM, Agustin GL, *et al*. Ubiquitous sensor networks in IMS: An ambient intelligence Telco platform. In: ICT Mobile Summit; 2008. p. 10–12.

[217] Dougherty D, Dunne D. Organizing ecologies of complex innovation. Organ Science. 2011;22(5).

[218] Gawer A. In: Gawer A, editor. Platform dynamics and strategies: From products to services. Surrey: Edward Elgar Publishing; 2009.

[219] Gawer A, Cusumano MA. Platform leadership: How Intel, Microsoft, and Cisco drive industry innovation. Boston, MA: Harvard Business Press; 2002.

[220] Wheelwright S, Clark K. Creating project plans to focus product development. Harvard Business Review. 1992;70(2):70–82.

[221] Meyer M, Lehnerd A. The Power of Product Platforms: Building Value and Cost Leadership. New York, NY: Free Press; 1997.

[222] McGrath M. Product Strategy for High Technology Companies (2nd ed.). New York, NY: McGraw-Hill; 2000.

[223] Krishnan V, Ulrich KT. Product development decisions: A review of the literature. Management Science. 2001;47(1):1–21.

[224] Panzar JC, Willig RD. Economies of scope. The American Economic Review. 1981;71(2):268–272.

[225] Parker GG, Alstyne MWV. Two-sided network effects: A theory of information product design. Management Science. 2005;51(10):1494–1504.

[226] Pine B. Mass customization: The new frontier in business competition. Boston, MA: Harvard Business School Press.; 1993.

[227] Suh DWO E, Chang D. Flexible product platforms: Framework and case study. Research Engineering Design. 2007;18(2):67–89.

[228] Bosch J. Maturity and Evolution in Software Product Lines: Approaches, Artefacts and Organization; 2002.

[229] Iansiti M, Levien R. The Keystone Advantage: What the New Dynamics of Business Ecosystems Mean for Strategy, Innovation, and Sustainability. Boston, MA: Harvard Business School Press; 2004.

[230]   Jansen BS S, Finkelstein A (2009) P. Business Network Management as a Survival Strategy: A Tale of Two Software Ecosystems; 2009.

[231]   Zirpoli F, Caputo M. The nature of buyer–supplier relationships in co-design activities. International Journal of Operations and Production Management. 2002;22(12):1389–1410.

[232]   Hammel T, Phelps T, Kuettner D. Re-engineering HP's CDRW supply chain. Supply Chain Management: An International Journal. 2002;7(3): 113–118.

[233]   Rochet JC, Tirole J. Platform competition in two-sided markets. Journal of the European Economic Association. 2003;1(4):990–1029.

[234]   Baldwin C, Woodard J. The Architecture of Platforms: A Unified View. Platforms, Markets and Innovation. London: Edward Elgar; 2009.

[235]   Bosch J. From Software Product Lines to Software Ecosystems. In: Proceedings of the 13th International Software Product Line Conference, San Francisco, CA; 2009.

[236]   Hagiu A. Multi-sided Platforms: From Microfoundations to Design and Expansion Strategies. Harvard Business School Working Papers; 2006.

[237]   Eisemann TR, Parker G, Van Alstyne M. In: Gawer A, editor. Opening Platforms: How, When and Why? Cheltenham: Edward Elgar Publishing Limited; 2009. p. 131–162.

[238]   Moore J. Predators and prey: A new ecology of competition. Harvard Business Review. 1993;71(3):75–83.

[239]   Peltoniemi M, Vuori E. Business ecosystem as the new approach to complex adaptive business environments. Power. 2004;p. 1–15.

[240]   Tiwana A, Konsynski B, Bush AA. Research commentary—Platform evolution: Coevolution of platform architecture, governance, and environmental dynamics. Info Sys Research. 2010;21(4):675–687.

[241]   Hanseth O, Lyytinen K. Design theory for dynamic complexity in information infrastructures: the case of building internet. Journal of Information Technology. 2010;25(1):1–19.

[242]   Baldwin C, Woodard J. The architecture of platforms: A unified view. In: A. Gawer (ed.), Platforms, Markets and Innovation. Cheltenham: Edward Elgar Publishing; 2009.

[243]   Boudreau K. The Boundaries of the Platform: Vertical Integration and Economic Incentives in Mobile Computing. Massachusetts Institute of Technology (MIT), Sloan School of Management Working papers. 2006.

[244]   Bharadwaj A, Sawy OAE, Pavlou PA, *et al.* Digital business strategy: Toward a next generation of insights. MIS Quarterly. 2013;37(2):471–482.

[245]   Tilson D, Lyytinen K, Sacrensen C. Research commentary digital infrastructures: The missing is research agenda. Information Systems Research. 2010;21(4):748–759.

[246]   Yoo Y, Henfridsson O, Lyytinen K. Research commentary: The new organizing logic of digital innovation: An agenda for information systems research. Information Systems Research. 2010;21(4):724–735.

[247] Schilling MA. Protecting or Diffusing a Technology Platform: Tradeoffs in Appropriability, Network Externalities, and Architectural Control. In: A. Gawer (ed.), Platforms, Markets and Innovation. Cheltenham: Edward Elgar Publishing; 2009.

[248] Gawer A, Cusumano M. How companies become platform leaders. MIT/Sloan Management Review. 2012;49(2):28.

[249] Hagel III BJ J, Davidson L. Shaping strategy in a world of constant disruption. Harvard Business Review. 2008;86(10):80–89.

[250] Iyer LCVN B. Managing in a small world ecosystem: Some lessons from the software sector. California Management Review. 2006;48(3).

[251] Gawer A, Henderson R. Platform owner entry and innovation in complementary markets: Evidence from Intel. Journal of Economics and Management Strategy. 2007;16(1):1–34.

[252] Jansen BS S, Finkelstein A. Providing transparency in the business of software: A modeling technique for software supply networks. In: Camarinha-Matos LM, Afsarmanesh H, Novais P, Analide C (eds.), Establishing the Foundation of Collaborative Networks. Boston, MA: Springer; 2007.

[253] Popp K, Meyer R. Profit from Software Ecosystems: Professional Edition, Business Models, Ecosystems and Partnerships in the Software Industries. Books on Demand GmbH; 2010.

[254] Goldsmith S, Crawford S. The Responsive City: Engaging Communities Through Data-Smart Governance. San Francisco: Jossey-Bass Wiley; 2014.

[255] Schaffers H, Komninos N, Pallot M, Trousse B, Nilsson M, Oliveira A. Smart cities and the future internet: Towards cooperation frameworks for open innovation. The Future Internet Lect. Notes Comput. Sci., 2011;6656:431–446.

[256] Bakry SH. "This is what modern deregulation looks like": Co-optation and contestation in the shaping of the UK's Open Government Data Initiative. The Journal of Community Informatics. 2012;8(2).

[257] Casadesus-Masanell R, Ricart JE. From strategy to business models and onto tactics. Long Range Planning. 2010;43(2-3):195–215.

[258] Chesbrough HW. Business model innovation: Opportunities and barriers. Long Range Planning. 2010;43(2/3):354–363.

[259] Teece DJ. Business models, business strategy and innovation. Long Range Planning. 2010;43(2-3):172–194.

[260] Al-Debei MM, Avison D. Developing a unified framework of the business model concept. European Journal of Information Systems. 2010;19(3):359–376.

[261] Breuer H. Lean venturing: Learning to create new business through exploration, elaboration, evaluation, experimentation, and evolution. International Journal of Innovation Management. 2013;17(03):1340013.

[262] Morris M, Schindehutte M, Allen J. The entrepreneur's business model: Toward a unified perspective. Journal of Business Research. 2005;58(6):726–735.

[263] Wirtz B. Business Model Management: Design, Instruments, Success Factors. Wiesbaden: Gabler; 2011.

[264] Geels BF, Elzen, Green K. In: Elzen F B, Geels, Green K, editors. General Introduction: System Innovation and Transitions to Sustainability. Cheltenham: Edward Elgar; 2004.

[265] Hockerts K, Wastenhagen R. Greening Goliaths versus emerging Davids Theorizing about the role of incumbents and new entrants in sustainable entrepreneurship. Journal of Business Venturing. 2010;25(5):481–492.

[266] Schaltegger S, Wagner M. Sustainable entrepreneurship and sustainability innovation: Categories and interactions. Business Strategy and the Environment. 2011;20(4):222–237.

[267] Hall J, Wagner M. Integrating sustainability into firms' processes: Performance effects and the moderating role of business models and innovation. Business Strategy and the Environment. 2012;21(3):183–196.

[268] Rayport JF, Sviokla J. Exploring the virtual value chain. Harvard Business Review. 1995;73(6):75–85.

[269] Bouwman H, van den Hooff B, van de Wijngaert L, van Dijk JAGM. Information and Communication Technology in Organizations. London: Sage; 2005.

[270] Bouwman H, de Vos H, Haaker T. Mobile Service Innovation and Business Models. Berlin: Springer; 2008.

[271] Matthing J, Kristensson P, Gustafsson A, *et al.* Developing successful technology based services: The issue of identifying and involving innovative users. Journal of Services Marketing. 2006;20(5):288–297.

[272] Hertog PD. Knowledge-intensive business services as co-producers of innovation. International Journal of Innovation Management. 2000;04(04): 491–528.

[273] Berry SVPJCS L, Dotzel T. Creating new markets through service innovation. MIT Sloan Management Review. 2006;(Winter):56–63.

[274] Chesbrough H, Rosenbloom RS. The role of the business model in capturing value from innovation: evidence from Xerox Corporation's technology spin off companies. Industrial and Corporate Change. 2002;11(3):529–555.

[275] Dubosson-Torbay OAM, Pigneur Y. E-business model design, classification, and measurements. Thunderbird International Business Review. 2002;44(1):5–23.

[276] Timmers P. Business models for electronic markets. Journal on Electronic Markets. 1998;8(2):3–8.

[277] Osterwalder Y, Alexander P, Tucci CL. Clarifying business models: Origins, present, and future of the concept. Communications of the Association for Information Systems. 2005;16(1).

[278] Amit R, Zott C. Value creation in e-business. Strategic Management Journal. 2001;22:493–520.

[279] Hawkins R. The phantom of the marketplace: Searching for new E-commerce business models. Communication and Strategies. 2002;6(2):297–329.

[280]   Mahadevan B. Business models for internet based E-commerce. California Management Review. 2000;42(4):55–69.

[281]   Tapscott LAD, Ticoll D. Digital Capital – Harnessing the Power of Business Webs. Boston: Harvard Business School Press; 2000.

[282]   Ajit K, Van Heck E. Making Markets. Boston: Harvard Business School Press; 2002.

[283]   Hedman J, Kalling T. The business model concept: Theoretical underpinnings and empirical illustrations. European Journal of Information Systems. 2003;12(1):49–59.

[284]   Porter M. Strategy and the internet. Harvard Business Review. 2001;(March): 63–76.

[285]   Hawkins R. 4. In: Preissl B, Bouwman H, Steinfield C, editors. Looking Beyond the Dot Com Bubble: Exploring the Form and Function of Business Models in the Electronic Marketplace. Heidelberg: Physica-Verlag HD; 2004. p. 65–81.

[286]   Johnson MW, Christensen CM, Kagermann H. Reinventing Your Business Model. Harvard Business Review. 2008;86:50–59.

[287]   Janssen M, Kuk G, Wagenaar RW. A survey of Web-based business models for e-government in the Netherlands. Government Information Quarterly. 2008;25(2):202–220.

[288]   Alt R, Zimmerman H. Introduction to special section Business models. Electronic Markets. 2001;11(1):3–9.

[289]   Daft RL, Lewin AY. Where are the theories for the "new" organizational forms? An editorial essay. Organization Science. 1993;4(4):1–6.

[290]   Dunbar RLM, Starbuck WH. Learning to design organizations and learning from designing them. Organization Science. 2006;17(2):171–178.

[291]   Mendelson H. Organizational architecture and success in the information technology industry. Management Science. 2000;46(4):513–529.

[292]   Brynjolfsson E, Hitt L. In: W Dutton RO B Kahin, Wyckoff A, editors. Intangible Assets and the Economic Impact of Computers. Boston, MA: MIT Press; 2004. p. 27–48.

[293]   Ghaziani A, Ventresca MJ. Keywords and cultural change: Frame analysis of business model public talk 1975-2000. Sociological Forum. 2005;20: 523–559.

[294]   Magretta J. Why Business Models Matter. Harvard Business Review. 2002;80:3–8.

[295]   Yip GS. Using strategy to change your business model. Business Strategy Review. 2004;15(2):17–24.

[296]   Shafer HS, Smith, Linder J. The power of business models. Business Horizons. 2005;48(3):199–207.

[297]   Afuah A, Tucci CL. Internet Business Models and Strategies: Text and Cases. New York: McGraw-Hill; 2001.

[298]   Weill P, Vitale MR. Place to Space: Migrating to E-business Models. Boston, MA: Harvard Business School Press; 2001.

[299]   Applegate LM. Emerging E-business Models: Lessons from the Field. Cambridge, MA: Harvard Business School; 2001.

[300]   Brousseau PT E. The economics of digital business models: A framework for analyzing the economics of platforms. Review of Network Economics. 2006;6(2):81–110.

[301]   Hazlett SA, Hill F. E-government: The realities of using IT to transform the public sector. Managing Service Quality: An International Journal. 2003;13(6):445–452.

[302]   Bakry SH. Development of e-government: A STOPE view. International Journal of Network Management. 2004;14(5):339–350.

[303]   Esteves J, Joseph RC. A comprehensive framework for the assessment of eGovernment projects. Government Information Quarterly. 2008;25(1): 118–132.

[304]   Ballon P. Business modelling revisited: The configuration of control and value. Journal of policy, regulation and strategy for telecommunications, Information and Media. 2007;(9):6–19.

[305]   Walravens N. Case study validation of a business model framework for smart city services: FixMyStreet and London bike app. IT Convergence Practice. 1992;(1):22–38.

[306]   Hamel G. Leading the Revolution. Boston: Harvard Business School Press; 2000.

[307]   Brandenburger AM, Nalebuff BJ. Co-Opetition. Benton Harbor, MI: Currency Doubleday; 1997.

[308]   Northrop EA L. Ultra-Large-Scale Systems The Software Challenge of the Future. Pittsburgh: Software Engineering Institute; 2006.

[309]   van Lamsweerde A. Goal-oriented Requirements Engineering: A Guided Tour. In: Proceedings of the Fifth IEEE International Symposium on Requirements Engineering, Toronto, Canada; 2001. p. 249–262.

[310]   Basole R. Visualization of interfirm relations in a converging mobile ecosystem. Journal of Information Technology. 2009;24:144–159.

[311]   De Reuver M, Bouwman H. Governance mechanisms for mobile service innovation in value networks. Journal of Business Research. 2012;65(3):347–354.

[312]   Hoffmann NK W H, Speckbacher G. The effect of interorganizational trust on make or cooperate decisions: Disentangling opportunism dependent and opportunism independent effects of trust. European Management Review. 2010;7(2):101–115.

[313]   Volz PBSC D, Anderl R. International Journal of Product Development. 2011;15(1):71–89. 10.1504/IJPD.2011.043662.

[314]   Chen Z, Dubinsky A. A conceptual model of perceived customer value in E-commerce: A preliminary investigation. Psychology and Marketing. 2003;20(4):323–347.

[315]   Kotler P. Marketing Management: Analysis, planning, implementation, and control. Englewood Cliffs, NJ: Prentice-Hall; 1988.

[316]   Pine J, Gilmore J. The Experience Economy. Boston: Harvard Business School Press; 1999.

[317] Davies T. Open Data, Democracy and Public Sector Reform: A Look at Open Government Data Use at data.gov.uk; 2010.

[318] Rogers EM. Diffusion of Innovations. New York: Free Press; 1995.

[319] Ilie V, Slyke CV, Green G, *et al.* Gender differences in Perceptions and use of communication technologies: A diffusion of innovation approach. Information Resources Management Journal. 2005;18(3):13–31.

[320] Bouwman H, de Reuver M, Dröes RM, *et al.* Creating Successful ICT-services. Practical Guidelines based on the STOF Method. Frux Freeband; 2008.

[321] Porter ME. Competitive Advantage: Creating and Sustaining Superior Performance. Free Press; 1998.

[322] Kotler P, Armstrong G, Harris LC, Piercy N. Principles of Marketing: The European Edition. Hemel Hempstead: Prentice-Hall; 1996.

[323] Harvey D. The right to the city. International journal of urban and regional research. 2003;27(4):939–941.

[324] Bowker SL G. Sorting Things Out: Classification and Its Consequences. Cambridge: MIT Press; 1999.

[325] Gitelman L. Data Bite Man: The Work of Sustaining Long-term Study. Cambridge: MIT Press.

[326] Glasmeier A, Christopherson S. Thinking about smart cities. Cambridge Journal of Regions, Economy and Society. 2015;8(1):3–12.

[327] Offenhuber D. Infrastructure legibility – A comparative analysis of open311-based citizen feedback systems. Cambridge Journal of Regions, Economy and Society. 2015;8(1):93–112.

[328] Crawford K, Schultz J. Big data and due process: Toward a framework to redress predictive privacy harms. Boston College Law Review. 2014;55(1):93–128.

[329] Joh EE. Policing by numbers: Big data and the fourth amendment. Washington Law Review. 2014;89:35–68.

[330] Ekbia H, Mattioli M, Kouper I, *et al.* Big data, bigger dilemmas: A critical review. Journal of the Association for Information Science and Technology. 2015;66.

[331] Carvalho L. Smart cities from scratch? A socio-technical perspective. Cambridge Journal of Regions, Economy and Society. 2014;8(1):43–60.

[332] Böhme R, Christin N, Edelman B, Moore T. Bitcoin: Economics, technology, governance. Journal of Economic Perspectives. 2015;29:213–38.

[333] Tapscott D, Tapscott A. Blockchain Revolution. New York: Penguin; 2016.

[334] P J Kirs KP, Kroeck G. A process model cognitive biasing effects in information systems development and usage. Information & Management. 2001;38(3):153–165.

[335] Tversky A, Kahneman D. Judgment under uncertainty: Heuristics and biases. Science. 2010;1974(4157).

[336] Zhang Q, Cheng L, Boutaba R. Cloud computing: State-of-the-art and research challenges. Journal of Internet Services and Applications. 2010;1: 7–18.

[337]   Villari M, Brandic I, Tusa F. Achieving Federated and Self-Manageable Cloud Infrastructures: Theory and Practice. IGI Global; 2012.

[338]   Townsend A. Smart Cities: Big Data, Civic Hackers, and the Quest for a New Utopia. New York: W.W. Norton & Co; 2013.

[339]   Zook M. Code/Space: Software and Everyday Life. Cambridge, MA: MIT Press; 2011.

[340]   Ramez SN, Elmasru E. Fundamentals of Database Systems. 6th ed. Harlow: Addison Wesley; 2010.

[341]   Kent W. Data and Reality. 1st Books Library; 2000.

[342]   Lancy D. 3D Data Management Controlling Data Volume Velocity and Variety. Stamford, CT: Meta Group – Gartner Research; 2001.

[343]   Rundel PW, Graham EA, Allen MF, *et al.* Environmental sensor networks in ecological research. New Phytologist. 2009;182(3):589–607.

[344]   Expert P, Evans T, Blondel V, *et al.* Uncovering Space-Independent Communities in Spatial Networks. In: Proceedings of the National Academy of Sciences of the United States of America 108; 2011. p. 7663–7668.

[345]   Scellato S, Noulas A, Lambiotte R, *et al.* Socio-spatial Properties of Online Location-based Social Networks. In: Proceedings of Fifth International AAAI Conference on Weblogs and Social Media; 2011.

[346]   McDaniel P, McLaughlin S. Security and privacy challenges in the smart grid. IEEE Security and Privacy. 2009;7(3):75–77.

[347]   Terzis S, Nixon P, Narasimhan N, *et al.* Middleware for pervasive and ad hoc computing. Personal Ubiquitous Computing. 2005;10(1):4–6.

[348]   Bahsoon R, Emmerich W. ArchOptions: A Real Options-Based Model for Predicting the Stability of Software Architecture. In: Fifth Workshop on Economics-Driven Software Engineering Research, EDSER 5, held in conjunction with the 25th International Conference on Software Engineering, Portland, OR; 2003.

[349]   Korn N, Oppenheim C. Licensing Open Data: A Practical Guide; 2011. Available from: http://discovery.ac.uk/files/pdf/Licensing_Open_Data_A_Practical_Guide.pdf.

[350]   Zappia Z. Participation, Power, Provenance: Mapping Information Flows in Open Data Development. Oxford Internet Institute; 2011. Available at https://zarino.co.uk/media/thesis.pdf.

[351]   Gurstein MB. Open data: Empowering the empowered or effective data use for everyone? First Monday. 2011;16(2).

[352]   Koppenjan GJ J. Institutional design for complex technological systems. International Journal of Technology. 2005;5(3):240–257.

[353]   Dawes SS, Cresswell AM, Pardo TA. From need to know to need to share: Tangled problems, information boundaries, and the building of public sector knowledge networks. Public Administration Review. 2009;69(3): 392–402.

[354]   Rittel H, Webber M. Dilemmas in a general theory of planning. Policy Sciences. 1973;4(2):155–169.

[355] Weber EP, Khademian AM. Wicked problems, knowledge challenges, and collaborative capacity builders in network settings. Public Administration Review. 2008;68(2):334–349.

[356] Selz D. Value Webs, Emerging forms of Fluid and Flexible Organizations: Thinking, Organizing, Communicating, and Delivering Value on the Internet; 1999. St. Gallen University, Dissertation. http://digitool.hbz-nrw.de:1801/webclient/DeliveryManager?pid=1136601andcustom_att_2=simple_viewer.

[357] Powell MW. Neither market nor hierarchy: Network forms of organisation. Research in Organisational Behaviour. 1990;12:295–336.

[358] Galbreath J. Twenty first century management rules: The management of relationships as intangible assets. Management Decision. 2002;40(2): 116–126.

[359] Hagel III J, AA. Net Gain: Expanding Markets through Virtual Communities. Boston, FL: Harvard Business School Press; 1997.

[360] Klein-Woolthuis R. Sleeping with the Enemy, Trust and Dependence in Interorganisational Relationships. Enschede: University of Twente; 1999.

[361] Stabell CB, Fjeldstad D. Configuring value for competitive advantage: On chains, shops, and networks. Strategic Management Journal. 1998;19(5): 413–437.

[362] Guiltinan JP. The price bundling of services: A normative framework. Journal of Marketing. 1987;51(2):74–85.

[363] Demkes R. COMET: A comprehensive methodology for supporting telematics investment decisions. Enschede: Telematica Instituut; 1999.

[364] Miller R, Lessard D. The Strategic Management of Large Engineering Projects. Shaping Institutions, Risks and Governance. Boston: MIT Press; 2000.

[365] Fong S, Meng HS. A Web-based Performance Monitoring System for E-government Services. In: 3rd International Conference on Theory and Practice of Electronic Governance. Bogota, Colombia, ACM; 2009.

[366] Boudreau K. Open platform strategies and innovation: Granting access vs. devolving control. Management Science. 2010;56(10):1849–1872.

[367] Acar AJM S, Novak K. Improving Access to Government through Better Use of the Web. W3C Interest Group Note; 2009.

[368] Anderson C. The Long Tail: How Endless Choice Is Creating Unlimited Demand. Random House; 2007.

[369] Bennett D, Harvey A. Publishing Open Government Data. W3C Working Draft; 2009.

[370] Chopra S, Meindl P. Supply Chain Management. Strategy, Planning and Operation. 5th ed. Harlow: Pearson Education; 2012.

[371] Kranton RE, Minehart DF. A theory of buyer-seller networks. American Economic Review. 2001;91(3):485–508.

[372] Generous N, Fairchild G, Deshpande A, Del Valle SY, Priedhorsky R. Global disease monitoring and forecasting with Wikipedia. PLOS Computational Biology. 2014;10(11):1–16.

[373]   Cachon GP, Fisher M. Supply chain inventory management and the value of shared information. Management Science. 2000;46(8):1032–1048.

[374]   Fisher ML. What is the right supply chain for your product? Harvard Business Review. 1997;75:105–117.

[375]   Girardin F, Calabrese F, Fiore FD, *et al.* Digital footprinting: Uncovering tourists with user-generated content. IEEE Pervasive Computing. 2008;7(4): 36–43.

[376]   Defee CC, Esper T, Mollenkopf D. Leveraging closed-loop orientation and leadership for environmental sustainability. Supply Chain Management. 2009;14(2):87–98. Cited By :30 Export Date: 14 May 2015.

[377]   Prahinski C, Kocabasoglu C. Empirical research opportunities in reverse supply chains. Omega. 2006;34(6):519–532.

[378]   Rogers DS, Tibben-Lembke R. An examination of reverse logistics practices. Journal of Business Logistics. 2001;22(2):129–148.

[379]   Thierry M, Salomon M, van Nunen J, *et al.* Strategic issues in product recovery management. California Management Review. 1995;37(2):114–135.

[380]   de la Fuente MV, Ros L, Card 3s M. Integrating forward and reverse supply chains: Application to a metal-mechanic company. International Journal of Production Economics. 2008;111(2):782–792.

[381]   Fleischmann M, Krikke HR, Dekker R, *et al.* A characterisation of logistics networks for product recovery. Omega. 2000;28(6):653–666.

[382]   Kleindorfer PR, Singhal K, Van Wassenhove LN. Sustainable operations management. Production and Operations Management. 2005;14(4):482–492.

[383]   Guide VDR, Jayaraman V, Srivastava R, *et al.* Supply-chain management for recoverable manufacturing systems. Interfaces. 2000;30(3):125–142.

[384]   Rashid A, Asif FMA, Krajnik P, *et al.* Resource conservative Manufacturing: An essential change in business and technology paradigm for sustainable manufacturing. Journal of Cleaner Production. 2013;57(0):166–177.

[385]   Vachon S, Klassen RD. Green project partnership in the supply chain: The case of the package printing industry. Journal of Cleaner Production. 2006;14 (6-7):661–671.

[386]   Krikke HR. Recovery Strategies and Reverse Logistic Network Design. In: Sarkis J, editor. Greener Manufacturing and Operations: From Design to Delivery and Back. Sheffield: Greenleaf Publishing; 1998.

[387]   Porter ME, Kramer MR. Strategy and society: The link between corporate social responsibility and competitive advantage. Harvard Business Review. 2006;84:78–92.

[388]   Ferro E, Osella M. Business Models for PSI Re-Use: A Multidimensional Framework. In: Using Open Data: Policy Modeling, Citizen Empowerment, Data Journalism Workshop. Brussels: European Commission; 2012.

[389]   Anderson JC, Narus JA. Business marketing: Understand what customers value. Harvard Business Review. 1998;76.

[390]   Demil B, Lecocq X. Business model evolution: In search of dynamic consistency. Long Range Planning. 2010;3(2-3):227–246.

[391]   Foss NJ, Saebi T. Business models and business model innovation: Between wicked and paradigmatic problems. Long Range Planning. 2017;51(1):9–21.

[392]   Cortimiglia MN, Ghezzi A, Frank AG. Business model innovation and strategy making nexus: Evidence from a cross industry mixed methods study. R&D Management. 2016;46(3):414–432.

[393]   Mason H, Rohner T. The Venturing Imperative. Boston: Harvard Business Press; 2002.

[394]   EuropeanCommission. Public Procurement for Smart Cities. European Commission; 2013.

[395]   Hughes J, Lang KR, Vragov R. An analytical framework for evaluating peer-to-peer business models. Electronic Commerce and Applications. 2007;7(1):105–118.

[396]   Shadbolt N, O'Hara K, Berners-Lee T, *et al.* Linked open government data: Lessons from Data.gov.uk. IEEE Intelligent Systems. 2012;27(3):16–24.

[397]   Dimitrakos T. System Models, e-Risks and e-Trust. IFIP I3E. Zurich: Kluwer Academic Publishers; 2001.

[398]   Daignault MSea M. Enabling Trust Online; 2002.

[399]   Friedman PKea B. Trust online. Communications of the ACM. 2000;43(12).

[400]   McWilliam G. Building Stronger Brands through Online Communities. Sloan Management Review. 2000;p. 43–54.

[401]   Abernathy WJ, Utterback JM. Managing assets and skills: The key to a sustainable competitive advantage. California Management Review. 1989;31(2): 91–106.

[402]   Berthon P, Hulbert JM, Pitt LF. To serve or create? Strategic orientations toward customers and innovation. California Management Review. 1999;(42):7–58.

[403]   Kourtesis D, Alvarez-Rodriguez JM, Paraskakis I. Semantic-based QoS management in cloud systems: Current status and future challenges. Future Generation Computer Systems. 2014;32(0):307–323.

[404]   Adner R. The Wide Lens. New York: Portfolio; 2012.

[405]   Lu D. Shared network investment. Journal of Economics and Management Strategy. 2001;73(3).

[406]   Wallin J. Operationalizing Competencies. In: 5th Annual International Conference on Competence-Based Management. Helsinki; 2000.

[407]   Kotler P. Marketing management. International Edition – The Millennium Edition. Englewood Cliffs, NJ: Prentice-Hall; 2000.

[408]   Jenkins H. Convergence Culture: Where Old and New Media Collide. New York: NY: University Press; 2006.

[409]   Bogers AA M, Bastian B. Users as innovators: A review, critique, and future research directions. Journal of Management. 2010;36:857–875.

[410]   Grönroos C. From marketing mix to relationship marketing: Towards a paradigm shift in marketing. Management Decision. 1994;32(2):4–20.

[411]   Dvir R, Pasher E. Innovation engines for knowledge cities: An innovation ecology perspective. Journal of Knowledge Management. 2004;8(5):16–27.

[412]   Moser MA. What is smart about the smart communities movement? EJournal. 2001;10/11:1. Available from: http://www.ucalgary.ca/ejournal/archive/v10-11/v10-11n1Moser-print.html.

[413]   Nunes P, Johnson B. Stimulating Consumer Demand Through Meaningful Innovation. Accenture Institute for Strategic Change; 2002.

[414]   Chen EL, Ho KK. Demystifying innovation. Perspectives on Business Innovation: Management, Policy and Practice. 2002;8:46–52.

[415]   Chesbrough HW. Open Innovation: The New Imperative for Creating and Profiting from Technology. Boston: Harvard Business School Press; 2003.

[416]   Etzkowitz H. The Triple Helix of University – Industry – Government Implications for Policy and Evaluation. Science Policy Institute; 2002.

[417]   Cosgrave E, Arbuthnot K, Tryfonas T. Living labs, innovation districts and information marketplaces: A systems approach for smart cities. Procedia Computer Science. 2013;16:668–677.

[418]   Perkmann M, Walsh K. University industry relationships and open innovation: Towards a research agenda. International Journal of Management Reviews. 2007;9(4):259–280.

[419]   Hargadon A. Firms as knowledge brokers: Lessons in pursuing continuous innovation. California Management Review. 1998;40(3).

[420]   Feurstein K, Hribernik KA, Schumacher J, Hesmer A, Thoben K-D. Living labs: A new development strategy. In: Schumacher J and Niitamo VP, editor. European living labs: A new approach for human centric regional innovation. Berlin: Wissenschaftlicher; 2008.

[421]   Glaeser BCR E L. Why Are Smart Places Getting Smarter? Taubman Center Policy Briefs, PB-2006-2; 2006. Available from: http://www.hks.harvard.edu/rappaport/downloads/policybriefs/brief_divergence.pdf.

[422]   Bronstein Z. Industry and the smart city. Dissent. 2009;56(3):27–34.

[423]   Becker S, Brogi A, Gorton I, *et al.* Towards an engineering approach to component adaptation. Berlin, Heidelberg: Springer-Verlag; 2006.

[424]   Kaisler S, Armour F, Espinosa JA, Money W. Big data: Issues and challenges moving forward. In: Proceedings of the 46th IEEE Annual Hawaii international Conference on System Sciences (HICC 2013), Grand Wailea, Maui, Hawaii, 2013, p. 995–1004.

[425]   Khan A, Hornbæk K. Big data from the built environment. In: Proceedings of the 2nd International Workshop on Research in the Large. Association for Computing Machinery, 2011. p. 29–32.

[426]   Carroll JJ, Bizer C, Hayes P, Stickler P. Named graphs, provenance and trust. In: WWW '05 Proceedings of the 14th International Conference on World Wide Web, ACM, 2005, p. 613–622.

Index